William Cawthorne Unwin

The elements of machine design

William Cawthorne Unwin

The elements of machine design

ISBN/EAN: 9783337163877

Printed in Europe, USA, Canada, Australia, Japan

Cover: Foto ©Andreas Hilbeck / pixelio.de

More available books at **www.hansebooks.com**

THE ELEMENTS

OF

MACHINE DESIGN

Part II
CHIEFLY ON ENGINE DETAILS

BY

W. CAWTHORNE UNWIN, F.R.S.

B.SC., MEM. INST. CIVIL ENGINEERS
MEM. AM. PHILOSOPHICAL SOCIETY AND HON. MEM. FRANKLIN INST.
PROFESSOR OF ENGINEERING AT THE CENTRAL INSTITUTION OF THE CITY
AND GUILDS OF LONDON INSTITUTE, FORMERLY PROFESSOR OF
HYDRAULIC AND MECHANICAL ENGINEERING AT THE
ROYAL INDIAN ENGINEERING COLLEGE

NEW (THIRTEENTH) EDITION
REVISED AND ENLARGED (1891)

LONDON
LONGMANS, GREEN, AND CO.
AND NEW YORK
1895

PREFACE TO PART II

THE present volume contains Chapters VI. and XII. to XVII. of the older editions of 'Machine Design,' but these chapters have been revised throughout and more than half the present volume is entirely new. It deals chiefly, though not exclusively, with steam-engine details. The subjects discussed form, it is believed, a definitely related group, and involve the application of mechanical principles to cases on the whole more complex than those examined in the first part of this treatise.

The general data of the strength of materials and the laws relating to the dependence of dimensions on straining actions are given in Part I. But in the cases to which these laws are applied in Part I. the straining actions are generally steady or statical, or may for practical purposes be treated as being so. In the design of Transmissive Machinery, which is treated of in Part I., in addition to questions of strength, kinematical problems are those most commonly presented. In Part II., which relates to machines the object of which is to supply mechanical energy for doing work, the parts of which move at varying speeds, and are subjected to varying forces of inertia which are too large to be neglected, dynamical principles find a more frequent application.

A treatise such as this has a double function. It is

partly a collection of designs, data, and rules, of use in the workshop and engineering drawing-office. But it differs from mere collections of precedents or pocket-books in this, that the reasons for different arrangements or proportions are, as far as possible, discussed in connection with the examples given. In the less complicated cases fairly definite rules can be given based on scientific principles, as a guide to the practical engineer. In the more complicated cases, where conditions are too indefinite for any precise theory, at least an indication can be given of the extent to which approximate solutions are to be distrusted. It is not the object of machine science to reduce every problem of construction to a rule of thumb. Rather it furnishes limits within which a free judgment, based on experience, has to be used and alternatives amongst which choice is to be made.

It is another function of such a book as this to serve as a text-book for engineering students, furnishing a series of practical problems. By working through such problems a student becomes acquainted with the various requirements involved in machine construction, and the considerations which guide an engineer in designing machines.

It is with a view to students that some problems are treated here in a fairly complete and systematic way, although necessarily with less completeness than in special treatises on particular parts of engineering. Such special treatises are generally too detailed to be convenient for students' use, and to understand them a larger knowledge of practical details is required than a student has time to master. It was specially from an experience of the want of some account of valve gears better suited to students than the large and admirable monographs already published that the author was led to extend the chapters on that subject.

Preface

The best special monographs are intended to explain the use of a single method in solving all problems. But for a student's purpose an acquaintance with different methods is desirable. No one method is equally convenient in all cases, and initially it is useful to look at the same problem from different points of view.

It may be useful to repeat here that, where there is no express statement to the contrary, the units adopted are as follows :—

Dimensions in inches.

Loads or forces in lbs.

Stresses in lbs. per sq. in.

Velocities in feet per second; accelerations in feet per second, per second.

Work in foot lbs.

Speeds in revolutions per minute; angular velocities in radians per second.

Statical moments in inch lbs.

The subject of Machine Designing is so extensive that a very much larger treatise than this would be required for its full explanation. All that the author could hope to do was to make a reasonably careful selection of what to include and what to reject. Shortcomings there must be. It can only be hoped that, taking into account the limitations imposed, they are not excessively numerous or serious.

January 1891.

CONTENTS

CHAPTER I

ON PIPES AND CYLINDERS 1–27

CAST-IRON PIPES, 1; thickness of water mains, 2; velocity of flow and loss of head in mains, 6; thickness of steam pipes and steam cylinders of cast iron, 8; WROUGHT-IRON AND STEEL PIPES, 9; pipes of other materials, 13; wrought-iron and steel riveted pipes, 13; PIPE JOINTS, 14; flanged joints, 14; diagram of pipe proportions, 15; special forms of pipe joints, 19; joints for hydraulic mains, 19; cylinders for great internal pressure, 20; socket joints, 21; joints for lead pipes, 23; joints for wrought-iron pipes, 24; boiler tubes, 26; condenser tubes, 27

CHAPTER II

ARRANGEMENT AND PROPORTIONS OF STEAM-ENGINE CYLINDERS 28–52

Arrangement of mechanism, 28; combined and compound engine arrangements, 29; working steam pressures, 31; piston speed, 32; ratios of cylinder volumes, 32; CONSTRUCTION AND PROPORTIONS OF STEAM-ENGINE CYLINDERS, 33; clearance, 36; hydraulic press cylinders, 38; INDICATOR DIAGRAM, single-cylinder engine, 39; provisional determination of mean effective pressure in a compound engine, 42; more accurate construction of the indicator diagram—Case I. Woolf engine, 45; Case II. Receiver engine, 49

CHAPTER III

LINKWORK. CRANKS AND ECCENTRICS . . 53–98

Kinematic pairs and chains, 53; piston displacement, 56; piston or crosshead velocity, 58; curves of piston velocity, 58; forces acting at crosshead and crank-pin, 59; approximate treatment of the forces on crank-pin, 60; correction of approximate diagram of piston and crank-pin effort for inertia, 62; forces acting at the crank-pin when the steam pressure varies and the obliquity of connecting rod is taken into account, 64; curve of force due to inertia when the obliquity of connecting rod is taken into account, 67; exact determination of the influence of inertia on diagram of crank-pin effort, 68; graphic construction of curve of forces due to translated inertia of reciprocating parts, 72; determination of curve of crank-pin effort or twisting moment when the indicator diagram of the engine is given, 75; combination of effort curves for two or more engines, 77; direct influence of weight of reciprocating parts in vertical engines, 79; advantages and disadvantages of the inertia of reciprocating parts, 79; HAND LEVERS AND WINCH HANDLES, 81; ENGINE CRANKS, 84; straining action on crank arm, 84; strength of crank, 85; proportions of cranks, 88; built-up steel cranks, 91; ECCENTRICS, 92; width of bearing surface of eccentric sheave, 92; friction of slide valve, 93; radius of eccentric, 94; proportions of sheave, 95; proportions of strap, 96; proportions of rod, 97; friction of eccentric, 98

CHAPTER IV

CONNECTING RODS 99–120

Straining forces on connecting rods, 100; resistance of connecting rod to tension, 102; stability of connecting rods as columns, 102; check on calculation of rod taking into account forces due to inertia, 105; CONNECTING-ROD ENDS, 108; proportions of steps, 108; strength of ends, 108; forms of connecting-rod ends, 110; box ends, 113; marine connecting-rod end, 114; coupling-rod joint, 119

CHAPTER V

CROSSHEADS AND SLIDES 121–135

Forces acting at the crosshead, 122; crosshead pin, 123; forms of crosshead, 124; SLIDE BARS, 130; wearing surface of slide bars and slide blocks, 133; strength of slide bar, 134

CHAPTER VI

PISTONS AND PISTON RODS 136–159

Volume swept through, 136; work and velocity of piston, 137; influence of weight of piston, 137; construction of pistons, 138; strength of pistons, 140; locomotive pistons, 142; stationary-engine pistons, 142; marine-engine pistons, 145; hydraulic pistons, 147; THEORY OF SPRING RINGS for pistons, 147; PISTON RODS, 155; strength of piston rod, 157; modes of fixing piston to piston rod, 158

CHAPTER VII

STUFFING-BOXES 160–168

Proportions of stuffing-box and gland, 161; Yarrow's stuffing-box, 162; stuffing-boxes with metallic packing, 163; packing for stuffing-boxes, 165; cup-leather packing, 165; friction of cup-leathers, 167

CHAPTER VIII

FLYWHEELS 169–191

Flywheel radius and speed, 169; DETERMINATION OF WEIGHT OF FLYWHEEL for given fluctuation of speed, 173; coefficient of fluctuation of energy, 173; periodical excess and deficiency of energy, 174; coefficient of fluctuation of speed, 178; weight of flywheel when coefficient of fluctuation of energy and speed are given, 178; CONSTRUCTION OF FLYWHEELS, 179; nave with rings shrunk on, 181; fastenings of rim segments, 182; STRENGTH OF FLYWHEELS, 183; stresses in homogeneous flywheel at uniform speed, 186

CHAPTER IX

VALVES, COCKS, AND SLIDE VALVES . . 192–218

AUTOMATIC VALVES, 193; flap or butterfly valves, 195; India-rubber disc valves, 196; lift or puppet valves, 198; pressure on valve seat, 199; MECHANICALLY CONTROLLED VALVES, 201; double beat or equilibrium valves, 202; Sir W. Thomson's valves, 203; controlled hydraulically moved valves, 206; SLIDING VALVES AND COCKS, 207; Dewrance's valves, 212; THEORY OF THE ACTION OF AUTOMATIC VALVES, 212; effective closing force of valves, 215; Prof. Riedler's valves, 217

CHAPTER X

VALVE GEARS 219–253

Steam-engine slide valves, 219; disturbance introduced by obliquity of eccentric rod, 261; setting valve to equalise lead, 221; lap and lead of slide valve, 221; area of steam ports, 224; proportions of slide valves, 225; travel of valve and corresponding crank angle, when obliquity of eccentric rod is neglected, 226; piston position for given crank angles, 229; crank angles for given ratios of expansion, 229; graphic methods of determining the relation between piston travel and crank angle—Muller circles, 229; VALVE DIAGRAMS, 232; valve ellipses, 233; REULEAUX, REECH, OR COSTE AND MANIQUET DIAGRAM, 236; exact diagram taking account of obliquity of eccentric and connecting rods, 236; complete valve diagram both for long and short eccentric rods, 238; problems, 241; ZEUNER'S POLAR VALVE DIAGRAM for a simple valve, 242; Theorem I. The polar locus of the valve travel reckoned from mid stroke is a pair of circles, 244; Zeuner's diagram with lap circles and Müller circles added, 245; CONSTRUCTION OF VALVES, 248; Trick valve, 249; balanced slide valve, 252

CHAPTER XI

EXPANSION VALVES AND LINK MOTIONS . . 254-282

Simple expansion valve, 254; action of expansion plate cutting off at outside edges, 255; MEYER VARIABLE EXPANSION GEAR, 258; complete diagram for a Meyer gear, 260; application of Zeuner's polar diagram to Meyer's valve gear, 262; complete polar diagram for a Meyer gear, 265; EXPANSION BY MOVABLE ECCENTRIC, 269; LINK MOTIONS, 270; travel of slide valve driven obliquely by an eccentric, 271; Stephenson's link motion, 272; approximate designing of a Stephenson link motion, 274; determination graphically of the exact travel of the valve for any crank position, 276; example of a link-motion valve gear, 279

CHAPTER XII

LUBRICATORS 283-287

Lubricants, 283; cup lubricator, 284; displacement lubricator, 285; needle lubricator, 286; Stauffer lubricator, 287

INDEX 289

ADDENDUM

Page 176. When in a semi-revolution there are (fig. 108) four values of the excess or deficiency of energy, $\Delta E_1, -\Delta E_2, \Delta E_3, -\Delta E_4$, then the greatest of these numerically is to be taken in calculating $K = \Delta E/E$; or, if still greater, the algebraic sum of any three consecutive values, such as $\Delta E_1 - \Delta E_2 + \Delta E_3$ or $-\Delta E_2 + \Delta E_3 - \Delta E_4$. This is also the value to be taken for ΔE in § 113

ELEMENTS
OF
MACHINE DESIGN

PART II

CHAPTER I

ON PIPES AND CYLINDERS

PIPES and cylinders subjected to internal pressure form parts of many machines, and are extensively used in the conveyance of water, gas, steam, or oil. The proportions of these, and the modes of joining them, form the subject of the present chapter.

CAST-IRON PIPES

1. Cast-iron pipes appear to have been first made about 1780 at Coalbrookdale. They are made to be joined by flanges or by a socket, the latter being more easily cast and less costly. Ordinarily they are made in lengths of 9 feet up to 12 in. (internal) diameter, and 12 feet long when larger. They should be cast vertically to insure soundness. When used in large quantities they are carefully inspected and tested, and are expected to be uniform in thickness and of normal weight. A variation of thickness in different parts of more than $\frac{1}{8}$ in., or a variation of more than 2 to 5 per cent. from the normal weight, would cause their rejection. They are usually tested by water pressure to double their intended working pressure.

2. *Thickness of cast-iron pipes for water mains.*—The rule for the thickness of a cylindrical vessel necessary to resist an internal or bursting pressure is given in I. § 26.

Let t = thickness of cylinder in inches.
d = diameter   ,,   ,,
p = excess of internal over external pressure in lbs. per sq. in.
H = pressure measured in feet of water.
f = safe limit of stress in lbs. per sq. in.

Then $p = 0.4333$ H.
H $= 2.308\ p$.

$$t = \frac{p\,d}{2f} = 0.2166\,\frac{\text{H}\,d}{f} \qquad . \qquad . \quad (1)$$

The average tenacity of the cast iron used for pipes may be taken at 18,500 lbs. per sq. in. Taking the factor of safety at $3\frac{1}{3}$, the highest safe tension is 5,500 lbs. per sq. in. Allowance must, however, be made—(*a*) for the irregular thickness of cast-iron pipes, which are often slightly thinner on one side than on the other; (*b*) for stresses due to hydraulic shock in the pipe, and to bending in consequence of pressure of the earth above, or settlement of the earth beneath, the pipe. A sufficient allowance will be made if the pipe is calculated for three times the actual working pressure, or, what amounts to the same thing, if the limit of stress is taken at one-third the value given above. Hence, the apparent factor of safety for pipes is $3 \times 3\frac{1}{3} = 10$, and the greatest safe stress, due to the actual pressure in the pipe, is 1,850 lbs. per sq. in.

In the mains used for the conveyance of water, the external pressure is 1 atmosphere, or 33 ft. of water pressure, and the greatest internal pressure is generally less than 7 atmospheres, or 231 ft. of water. Hence, the excess of internal over external pressure may be taken at 6 atmospheres, or 90 lbs. per sq. in. Putting this value in the formula above, we get

$$t = \frac{90\,d}{2 \times 1850} = 0.0231\,d \quad . \quad . \quad . \quad (2)$$

Internal diameter of pipe in ins.

4	8	12	16	20	24	30	36	42

Thickness of pipe in ins.

0.0924	0.185	0.277	0.370	0.462	0.554	0.693	0.832	0.970

Thickness to nearest sixteenth of an inch

$\frac{1}{8}$	$\frac{3}{16}$	$\frac{5}{16}$	$\frac{3}{8}$	$\frac{1}{2}$	$\frac{9}{16}$	$\frac{11}{16}$	$\frac{13}{16}$	1

3. This table shows that some of the thicknesses given by the above rule, although ample margin of strength has been allowed, are so small that the pipes could not be cast with any certainty of success. Many years ago the following rule for the thickness of water mains was given by Mr. Hawksley :—

$$t = 0.18\sqrt{d}$$

That rule represents, very fairly, the least thickness which it is desirable to attempt to cast. The following rule agrees still better with practical experience. Let $t_{min.}$ be the least thickness which should be adopted for a cylindrical pipe casting, of ordinary length and of diameter d, in order that there may be no special difficulty in getting it cast. Then

$$t_{min.} = 0.11\sqrt{d} + 0.1 \quad . \quad . \quad . \quad (3)$$

Diameter of pipe in ins.

4	8	12	16	20	24	30	36	42	48	54	60

Least thickness of pipe in ins.

.320	.411	.481	.540	.592	.639	.703	.760	.813	.862	.908	.953

Thickness to nearest sixteenth of an inch

$\frac{3}{8}$	$\frac{7}{16}$	$\frac{1}{2}$	$\frac{9}{16}$	$\frac{5}{8}$	$\frac{11}{16}$	$\frac{11}{16}$	$\frac{3}{4}$	$\frac{13}{16}$	$\frac{7}{8}$	$\frac{15}{16}$	1

Pipes of the thicknesses here given will in general be

safe for pressures not exceeding 6 atmospheres, or 90 lbs. per sq. in., when under 20 ins. diameter, and for 5 atmospheres, or 75 lbs. per sq. in., when under 60 ins. diameter. When pipes are subjected to greater pressure, it is desirable to use the more exact formula given in I. § 26 (eq. 3) in calculating the thickness. Putting in that formula $f = 1850$, it becomes

$$\frac{t}{d} = \frac{1}{2}\left\{\sqrt{\frac{2775+p}{2775-2p}} - 1\right\} \quad . \quad . \quad (4)$$

From this formula the following table has been calculated:—

Excess of Internal over External Pressure		Ratio of thickness to diameter of pipe
In lbs. per square inch	In feet of head of water	$\frac{t}{d}$
75	173	·021
90	208	·026
105	242	·030
120	277	·035
135	311	·039
150	346	·044
165	381	·048
180	415	·053
195	450	·058
210	484	·063
225	519	·068
250	577	·077
275	634	·085
300	692	·095
350	808	·114
400	923	·134
450	1039	·156
500	1154	·179
750	1731	·332
1000	2308	·603

4. In the following table the second and third columns give the least thickness of pipe which it is practicable to cast (from eq. 3), or the thickness to be adopted when equations (1) or (4) give a less value. The other columns give thick-

Pipes and Cylinders.

Internal diameter of pipe in ins.	Least thickness of pipe by eq. (3) || Thickness necessary for strength, by eq. (4), for working internal pressures in lbs. per sq. in, and ft. of head amounting to								
	Exact	Correct to nearest sixteenth	75 lbs. 173 ft.	90 lbs. 208 ft.	105 lbs. 242 ft.	120 lbs. 277 ft.	150 lbs. 346 ft.	180 lbs. 415 ft.	210 lbs. 484 ft.	250 lbs. 577 ft.
2	·256	¼	...	...	...	...	...	...	...	...
3	·291	5⁄16	...	...	...	...	...	...	...	...
4	·320	⅜	...	...	...	...	...	...	...	·385
5	·346	,,	...	...	...	...	...	...	·378	·462
6	·369	,,	...	...	...	...	...	·424	·441	·539
7	·391	7⁄16	...	...	...	...	...	·477	·504	·616
8	·411	,,	...	...	...	...	·528	·530	·567	·693
9	·430	½	...	...	...	·560	·616	·636	·630	·770
10	·448	,,	...	...	...	·630	·704	·742	·756	·924
12	·481	9⁄16	...	...	·66	·700	·792	·848	·882	1·08
14	·512	,,	...	...	·72	·770	·880	·954	1·01	1·23
16	·540	,,	...	·780	·90	·840	·968	1·060	1·13	1·39
18	·567	5⁄8	...	·936	1·08	1·05	1·056	1·166	1·26	1·54
20	·592	,,	·882	1·092	1·26	1·26	1·320	1·272	1·39	1·69
22	·616	,,	1·008	1·248	1·44	1·47	1·584	1·590	1·51	1·85
24	·639	11⁄16	1·134	1·404	1·62	1·68	1·848	1·908	1·89	2·31
30	·702	¾	1·260	1·560	1·80	1·89	2·112	2·226	2·27	2·77
36	·760	13⁄16	1·512	1·872	2·16	2·10	2·376	2·544	2·65	3·23
42	·813	,,	1·764	2·184	2·52	2·52	2·640	2·862	3·02	3·70
48	·862	⅞				2·94	3·168	3·180	3·40	4·16
54	·909	15⁄16					3·696	3·816	3·78	4·62
60	·952	,,						4·452	4·54	5·54
72	1·033	1							5·29	6·47
84	1·108	1 1⁄16								

nesses calculated by equation (4). It should be remembered that in obtaining these thicknesses an allowance has been made for bending stress, and hence somewhat less thicknesses may be adopted in pipes so supported as to be protected from any bending. To convert feet of head of water into lbs. per sq. in., multiply by 0·4333. In water mains for towns, a thickness about 25 per cent. greater than that given in this table is often adopted in practice.

5. *Velocity of flow and loss of head in water mains.*—The diameter d of water-mains is determined with reference to the volume Q in cubic feet per second to be delivered. If v is the velocity of flow in feet per second,

$$Q = \frac{\pi}{4} d^2 v$$

where the diameter d is in feet. Generally in water mains the velocity does not exceed 3 feet, or at most 4 feet, per second, in order that the shocks due to the momentum of the water when its velocity changes may not be too serious. In cases where it is desirable to lose as little energy as possible, the velocity is restricted to $1\frac{1}{2}$ or 2 feet per second.

In a large number of cases it is important to determine what loss of head occurs in long water mains. The head lost in friction may be a loss of level or a loss of pressure, and in either case is most conveniently reckoned in feet of water. Let h be the head lost in feet, in a main of length l feet, when water flows through it with a velocity v in feet per second. The most exact relation between these quantities is—

$$\frac{h}{l} = \frac{m}{d^x} \cdot \frac{v^n}{2g} \quad . \quad . \quad . \quad . \quad (5)$$

where m, n, and x are constants depending on the roughness of surface of the pipe. The equation is, however, inconvenient for practical use. More commonly, therefore, engineers use the equation,

Pipes and Cylinders

$$\frac{h}{l} = \frac{k}{d} \cdot \frac{v^2}{2g} \quad . \quad . \quad . \quad (6)$$

where k has a considerable range of values in different cases. In 1886 the author communicated to *Industries* a very careful determination of the constants m, n, and x, for all the most trustworthy experiments on flow in pipes. From these values it was then possible to calculate a series of values of k in equation (6) for such cases as commonly occur in practice. With the values given in the following tables, the loss of head in pipes can be calculated much more accurately than with a constant value for k, and nearly as accurately as if the more awkward equation (5) were used.

Clean Wrought-Iron Pipes

When d in feet is	The value of k for the following velocities in feet per second is			
	1 to 2	2 to 3	3 to 4	4 to 5
0·5 to 0·75	·0230	·0200	·0184	·0172
0·75 to 1·0	·0214	·0186	·0171	·0160
1·0 to 1·5	·0199	·0173	·0159	·0149
1·5 to 2·0	·0185	·0161	·0148	·0139
2·0 to 3·0	·0172	·0150	·0137	·0129
3·0 to 4·0	·0160	·0139	·0128	·0120

Asphalted Cast-Iron Pipes

When d in feet is	The value of k for the following velocities in feet per second is			
	1 to 2	2 to 3	3 to 4	4 to 5
0·5 to 0·75	·0257	·0236	·0224	·0216
0·75 to 1·0	·0246	·0226	·0215	·0207
1·0 to 1·5	·0235	·0216	·0206	·0198
1·5 to 2·0	·0225	·0207	·0197	·0189
2·0 to 3·0	·0215	·0198	·0188	·0181
3·0 to 4·0	·0206	·0190	·0180	·0173

New Cast-Iron Pipes

When d is in feet	The value of k for the following velocities in feet per second is			
	1 to 2	2 to 3	3 to 4	4 to 5
0·5 to 0·75	·0230	·0224	·0220	·0217
0·75 to 1·0	·0216	·0210	·0206	·0204
1·0 to 1·5	·0204	·0199	·0195	·0193
1·5 to 2·0	·0193	·0188	·0184	·0182
2·0 to 3·0	·0182	·0177	·0174	·0172
3·0 to 4·0	·0172	·0167	·0164	·0162

Incrusted Cast-Iron Pipes

When d in feet is	k is for all velocities
0·5 to 0·75	·0475
0·75 to 1·0	·0450
1·0 to 1·5	·0426
1·5 to 2·0	·0403
2·0 to 3·0	·0381
3·0 to 4·0	·0361

At first sight the co-efficients do not seem to differ greatly; but, for each kind of pipe, k varies by at least 20 per cent. within the restricted limits chosen. Also the resistance for incrusted cast iron is quite double that of clean wrought iron, and for the larger pipes and higher velocities more than this.

6. *Thickness of steam pipes and steam cylinders of cast iron.*—This is determined in precisely the same way as the thickness of water mains. For steam pipes the thicknesses given in the preceding table will answer. For steam cylinders an allowance has to be made for re-boring. The cylinder thickness may be obtained from the following equation:—

$$t = \frac{pd}{3700} + c = ·00027\, pd + c \quad . \quad . \quad (7)$$

where c ranges from 0·5 to 0·75 in carefully constructed engines, and is as much as 1·0 in some cases.

Pipes and Cylinders

The friction of steam is much smaller than that of water, from its less density; hence, steam pipes are generally designed so that the velocity of the steam should not exceed about 100 feet per second.

Wrought-Iron and Steel Pipes

7. Wrought-iron welded pipes came into extensive use for the distribution of gas. A strip of wrought-iron plate heated to welding heat was bent to cylindrical form and then welded by hammers or rolls or by passing through dies. Such tubes are always weakest at the weld, and when the diameter exceeds 6 or 7 inches are difficult to manufacture.

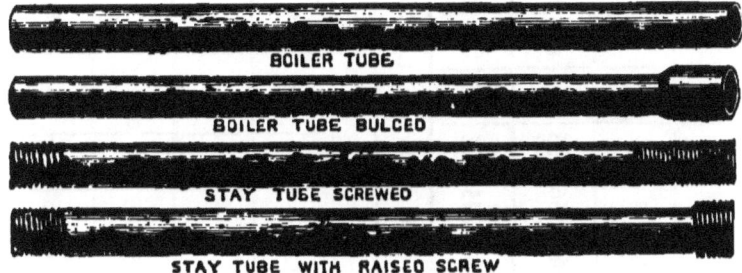

Fig. 1

For hydraulic purposes, where pressures of often 2 tons per sq. in. have to be transmitted, solid drawn weldless tubes have been produced in small sizes. Quite lately an entirely new process, invented by Messrs. Mannesmann, has been introduced, and seems likely to revolutionise the manufacture of tubes. It is not well adapted for wrought iron, but with ductile steel, copper, delta metal, or lead, weldless tubes are produced of almost any size by a peculiar process of rolling from a solid bar.[1]

Wrought-iron and steel pipes or tubes are obtainable of the following descriptions :—

(1) Butt-welded pipes used for gas, water, and steam,

[1] See a paper by Mr. J. G. Gordon, 'Journal of Society of Arts,' vol. xxxviii.

Wrought-Iron Tubes

Internal diameter in ins.	1	1¼	1½	1¾	2	2¼	2½	2¾	3	3¼	3½	3¾	4	4¼	4½	4¾	5	5¼
Thickness B. W. G.	9	9	9	9	8	8	8	8	7	7	7	7	7	6	6	6	6	5
Thickness in ins.	0·158	·158	·158	·158	·166	·166	·166	·166	·187	·187	·187	·187	·187	·208	·208	·208	·208	·217
Greatest working pressure in lbs. per sq. in.	1720	1370	1145	980	930	825	745	675	745	690	640	595	560	615	580	550	520	530
Greatest length							Usual length 12 to 14 feet. Can be made to 20 feet.											

Internal diameter in ins.	5½	5¾	6	6¼	6½	6¾	7	7½	8	8½	9	9½	10	10½	11	12
Thickness B. W. G.	5	5	5	4	4	4										
Thickness in ins.	·217	·217	·217	·239	·239	·239	0·25	0·25	0·25	0·3125	0·3125	0·3125	0·375	0·375	0·375	0·375
Greatest working pressure in lbs. per sq. in.	505	485	460	505	485	470	485	455	425	525	500	470	565	540	515	480
Greatest length	As above.									Length 16 or 17 feet.						

of wrought iron. These are made of from $\frac{1}{8}$ in. to 4 in. internal diameter, and in lengths usually not exceeding 14 feet. They can be obtained up to 20 feet in length if necessary. The steam tubes are two gauges thicker than the gas tubes.

(2) Lap-welded wrought-iron steam tubes, which are stronger than butt-welded tubes. They are proved to 400 lbs. per sq. in. before being sent out. The preceding table gives the usual dimensions of these tubes.

The thickness of such tubes will not be exactly regular. Suppose that when t is the nominal thickness, $t-\frac{1}{16}$ is the effective thickness which can be relied on in calculating the strength. Then by equation (2) I. § 26, for an internal bursting pressure,

$$f = \frac{pd}{2(t-\frac{1}{16})};$$

and taking the greatest safe stress at 4 tons or 8,960 lbs. per sq. in.

$$t = \cdot 0000558\, pd + \tfrac{1}{16} \quad . \quad . \quad (8)$$

By this rule the working pressures p for different diameters d, given in the table above, have been calculated. Lap-welded steam tubes of the dimensions given above are proved to 400 lbs. per sq. in. before being sent out.

Fig. 1 shows the ordinary forms in which wrought-iron and steel tubes can be obtained.

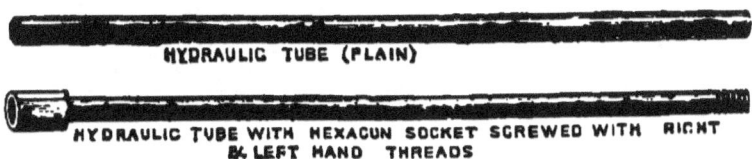

HYDRAULIC TUBE (PLAIN)

HYDRAULIC TUBE WITH HEXAGON SOCKET SCREWED WITH RIGHT & LEFT HAND THREADS

Fig. 2

(3) Weldless solid drawn steel tubes (fig. 2) are also manufactured, which are used for boiler tubes, hydraulic pipes, and occasionally for other purposes, such as hollow shafting and boring rods. The following tables give the usual dimensions of such tubes:—

Weldless Steel Boiler Tubes

External diameter in inches	$1\frac{1}{2}$	$1\frac{5}{8}$	$1\frac{3}{4}$	$1\frac{7}{8}$	2	$2\frac{1}{8}$	$2\frac{1}{4}$	$2\frac{3}{8}$	$2\frac{1}{2}$	$2\frac{5}{8}$	$2\frac{3}{4}$	$2\frac{7}{8}$	3	$3\frac{1}{8}$	$3\frac{1}{4}$	$3\frac{3}{8}$	$3\frac{1}{2}$
Thicknesses B. W. G.	16 to 10	16 to 10	16 to 10	16 to 10	16 to 10	15 to 10	15 to 10	14 to 10	14 to 10	14 to 10	14 to 10	13 to 10	13 to 10	13 to 10	13 to 10	13 to 10	13 to 10
Length in feet	13 feet.											10 feet.					

Weldless Steel Boiler and Hydraulic Tubes

External diameter in inches		$\frac{1}{2}$	$\frac{5}{8}$	$\frac{3}{4}$	$\frac{7}{8}$	1	$1\frac{1}{8}$	$1\frac{1}{4}$	$1\frac{3}{8}$	$1\frac{1}{2}$	$1\frac{5}{8}$	$1\frac{3}{4}$	$1\frac{7}{8}$	2
Thicknesses in inches	Least	$\frac{1}{16}$	$\frac{1}{16}$	$\frac{1}{16}$	$\frac{1}{16}$	$\frac{1}{16}$	$\frac{1}{16}$	$\frac{1}{16}$	$\frac{1}{16}$	$\frac{1}{16}$	$\frac{1}{16}$	$\frac{1}{16}$	$\frac{1}{16}$	$\frac{1}{16}$
	Greatest	$\frac{1}{8}$	$\frac{3}{16}$	$\frac{1}{4}$	$\frac{1}{4}$	$\frac{1}{4}$	$\frac{5}{16}$	$\frac{3}{8}$	$\frac{7}{16}$	$\frac{1}{2}$	$\frac{1}{2}$	$\frac{1}{2}$	$\frac{1}{2}$	$\frac{1}{2}$

External diameter in inches		$2\frac{1}{8}$	$2\frac{1}{4}$	$2\frac{3}{8}$	$2\frac{1}{2}$	$2\frac{5}{8}$	$2\frac{3}{4}$	$2\frac{7}{8}$	3	$3\frac{1}{8}$	$3\frac{1}{4}$	$3\frac{3}{8}$	$3\frac{1}{2}$
Thicknesses in inches	Least	$\frac{3}{32}$	$\frac{3}{32}$	$\frac{3}{32}$	$\frac{3}{32}$	$\frac{3}{32}$	$\frac{3}{32}$	$\frac{3}{32}$	$\frac{3}{32}$	$\frac{3}{32}$	$\frac{3}{32}$	$\frac{3}{32}$	$\frac{3}{32}$
	Greatest	$\frac{1}{2}$	$\frac{1}{2}$	$\frac{1}{2}$	$\frac{1}{2}$	$\frac{1}{2}$	$\frac{1}{2}$	$\frac{1}{2}$	$\frac{1}{2}$	$\frac{1}{2}$	$\frac{1}{2}$	$\frac{1}{2}$	$\frac{1}{2}$

Maximum length 15 feet.

The steel of which these tubes are made has a tensile strength of about 30 tons per sq. in. For hydraulic tubes the working strength may be taken at 10 tons per sq. in. Putting this for the value of f in equation (2), I. § 26, we get

$$t = 0\cdot 00002232\, p\, d, \qquad . \qquad . \qquad (9)$$

where p is the excess of internal over external pressure in lbs. per sq. in. and d the internal diameter of the tube.[1]

8. *Thickness of pipes of other materials:*

For lead pipes $\qquad t = \cdot 0025\, p\, d + \frac{3}{16}$
copper pipes $\qquad t = \cdot 00018\, p\, d + \frac{1}{16}$
wrought-iron pipes $\; t = \cdot 00006\, p\, d + \frac{1}{16}$ (if welded)
$\qquad\qquad\qquad\qquad = \cdot 00012\, p\, d + \frac{1}{16}$ (if riveted)[2]

Iron boiler tubes subjected to external pressure are $2\frac{1}{2}$ to 4 ins. diameter, and are strong enough for a pressure of 40 lbs. per sq. in. of the minimum thickness made. For other pressures

$$t = \cdot 00017\, p\, d + 0\cdot 1$$

9. *Copper steam pipes* have a thickness equal to $0\cdot 0001\, p\, d + \frac{1}{8}$ and copper feed pipes a thickness $0\cdot 00013\, p\, d + \frac{1}{8}$. In the older engines with moderate steam pressure the steam pipes were $\frac{1}{4}$ in. thick and the feed pipes $\frac{3}{16}$. Pipes of this kind are connected by tough brass flanges brazed to the pipes. If t is the thickness of the pipes, the flange thickness $= 3\,t$; bolt diameter at least $3\,t$; pitch of bolts not more than $15\,t$. Width of flanges $2\frac{1}{4}$ times the bolt diameter. The strength of bolts should be looked to.

10. *Wrought-iron and steel riveted water pipes.*—Water mains of comparatively large size of thin riveted wrought iron seem first to have been used in California in hydraulic mining. An account of these pipes will be found in a

---

[1] Messrs. E. Lewis and Sons of Wolverhampton, and the Weldless Steel Tube Company of Birmingham, supplied the information as to the ordinary dimensions of tubes manufactured.

[2] See, however, the more exact rules for riveted boilers in I. § 73.

paper by Mr. Hamilton Smith ('Trans. Am. Soc. of Civil Engineers, 1884'). These pipes are single or double riveted, and very often are put together stove-pipe fashion. That is, the pipe is made slightly conical; a piece of canvas is wrapped round the small end of one pipe and this is then forced into the large end of another by hydraulic jacks. Leakage is stopped if it occurs by small pine wedges. The pipes are sometimes only $\frac{1}{12}$ in. thick, and they are used under pressures which strain the metal at the riveted joint up to the singularly high-working stress of 16,000 to 18,000 lbs. per sq. in. They are protected against corrosion by a coating of tar, and no serious corrosion seems to occur. At bends they are strongly braced to prevent movement. More recently steel pipes of a similar kind have come into use for water supply where carriage is expensive. These are commonly double riveted, and the joints are made by patent sockets riveted to the pipe, or by rings of cast iron, forming a double socket.

Pipe Joints

11. Cast-iron pipes are connected by flange joints, or by spigot and faucet joints. The former are stronger, easier to connect or disconnect, and are always used when the pipes are placed vertically. The latter are less costly, and are better for pipes laid in the ground, because they permit the pipes to adapt themselves to the inequalities of the ground while being laid, and the line of pipes retains a slight flexibility.

12. *Flanged joints.*—The proportions of flanges have been to some extent given in I. § 85. Fig. 3 shows one form of flanged joint for pipes, for which the following proportions may be used:—

$$\text{Thickness } t_1 = t_2 = \tfrac{5}{4} t \text{ to } \tfrac{5}{4} t + \tfrac{1}{4}$$
$$t_3 = \tfrac{3}{2} t$$
$$t_4 = \tfrac{3}{8} \text{ in.}$$
$$\text{Width} = w = 2\,\hat{c} + \tfrac{1}{2} \text{ to } 2\,\hat{c} + \tfrac{3}{4}$$

Diam. of bolts $= \delta = 0.016\, d \sqrt{\dfrac{p}{n} + \dfrac{1.6}{d}}$

Number of bolts $= n = 2 + \dfrac{d}{2}$

Diam. of bolt-hole $= \delta + \tfrac{1}{8}$

The joint shown is made with a lead ring. The joint may be made by facing the flanges and bringing them together with a string smeared with red lead, or an india-rubber or gutta-percha ring interposed. A rough joint is made with a ring of wrought iron, covered with tarred rope, the space between the flanges being filled up with rust cement. For steam pipes a ring of asbestos millboard is often used.

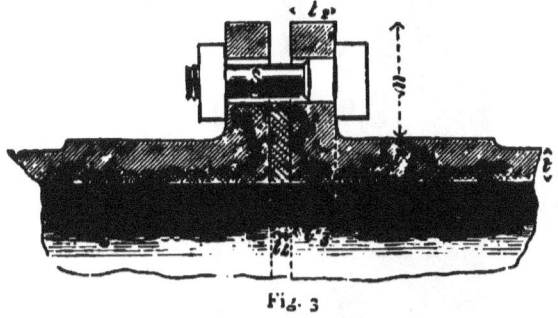

Fig. 3

Mr. Emery for steam pipes has used a ring of corrugated copper.

13. *Graphic diagram of pipe-joint proportions.*—Fig. 4 shows a very convenient graphic way of settling the proportions of simple machine parts for a series of different sizes. The curve or straight line which represents the proportions given by rule is first laid down as shown, for instance, by the dotted line ab. This is then broken up into a stepped line, the steps increasing by eighths of an inch or any other required difference. With a little dexterity such a diagram may be made to give all the required dimensions for any series of sizes without confusion. It is best drawn full size when that is possible. The method may be applied in many cases and is more convenient than a table.

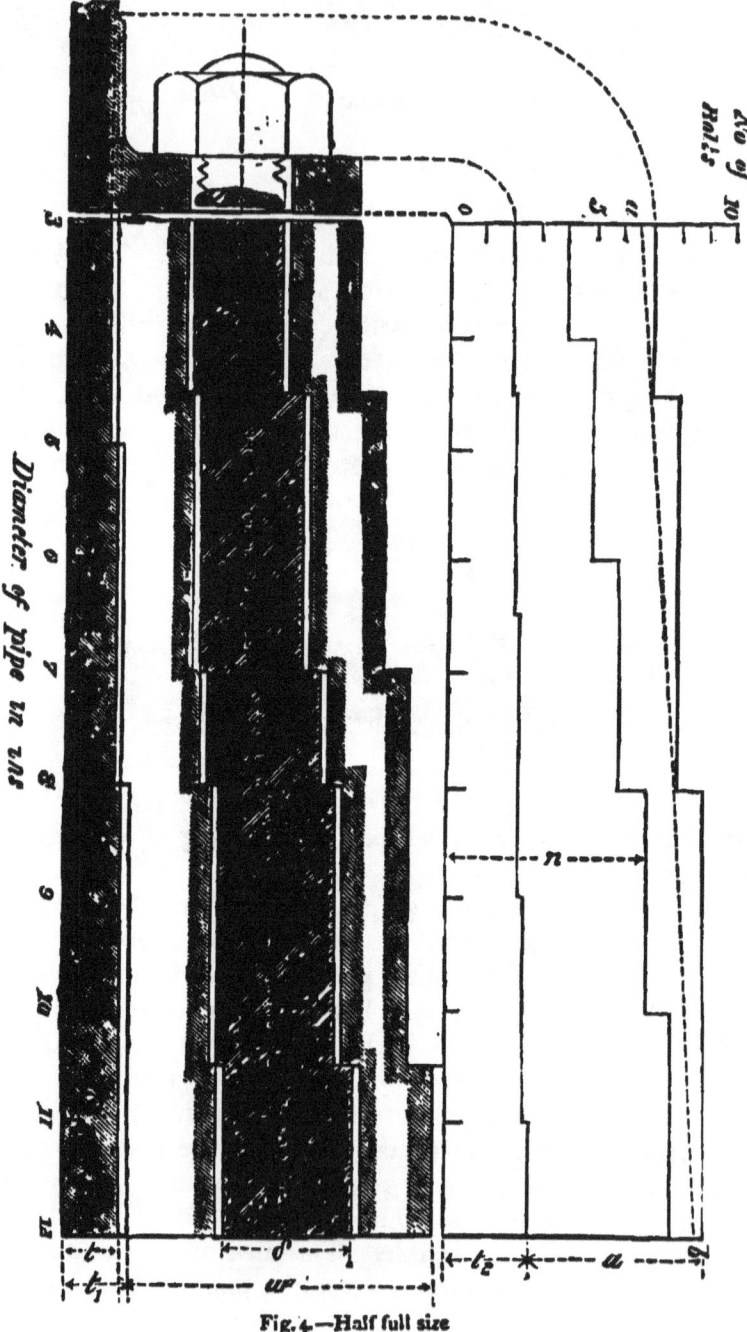

Fig. 4.—Half full size

Pipes and Cylinders

14. Mr. Robert Briggs laid down proportions for flanged pipes some years since which have been very widely adopted in American practice. As these have been tested by a large experience, it will be useful to quote them here for comparison with the rules given above. These rules are intended to apply (1) to pipes under pressures up to 75 lbs. per sq. in. (or 170 feet of head), the flanges being faced all over, and having bolts with hexagon nuts and heads; (2) to pipes for pressures up to 100 lbs. pressure per sq. in. (230 feet of head), with flanges thick enough to permit the use of packing rings placed within the bolts as a substitute for surfacing and wide enough for bolts with square nuts and heads. The form of Mr. Briggs's rules has been modified so as to render them applicable to pipes of thicknesses different from those given by him. Fig. 5 gives the reference symbols.

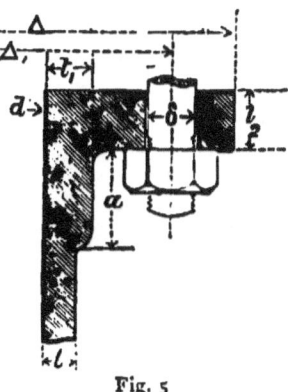

Fig. 5

Mr. Briggs's Rules for Pipe Flanges.

	Pipes for 75 lbs. pressure	Pipes for 100 lbs. pressure
Thickness of pipes $=t$	$=0.026d+0.25$	$0.0304d+0.3$
Thickness of boss $=t_1$	$= 1.2t$	$1.15t+0.1$
Length of boss $=a$	$= 1.92t+0.62$	$1.64t+0.61$
Thickness of flange (finished) $=$	$1.28t+0.08$	$1.31t+0.2$
" " (rough) $=$	$1.41t+0.10$	$1.48t+0.42$
Radius of hollow at angle	none	$0.26t+0.08$
Diameter of flange $=\Delta$	$=\left\{\begin{array}{l} d+2.4t+ \\ 4.28\delta+0.6 \end{array}\right\}$	$\left\{\begin{array}{l} d+2.82t+ \\ 5\delta+0.66 \end{array}\right\}$
Diameter of bolt circle $=\Delta_1$	$=\left\{\begin{array}{l} d+2.4t+ \\ 2.08\delta+0.4 \end{array}\right\}$	$\left\{\begin{array}{l} d+2.82t+ \\ 2.5\delta+0.56 \end{array}\right\}$
Diameter of holes in flange $=$	$1.03\delta+0.03$	$1.03\delta+0.03$
Number of bolts $=n$	$=0.705d+2.18$	$0.705d+2.18$

Diameter of bolts $=\delta=$

$$1{\cdot}182\sqrt{\left\{\frac{0{\cdot}01785\Delta_1^2+0{\cdot}2052}{n}\right\}}+0{\cdot}0492$$

in which equation we may take $\Delta_1=1{\cdot}1043d+2{\cdot}01$. The following rule is much simpler, and gives nearly the same values :—

$$\delta=\frac{0{\cdot}1674d+0{\cdot}305}{\sqrt{n}}+0{\cdot}05$$

The following table gives dimensions calculated by Mr. Briggs's rules :—

Cast-Iron Flanged Pipes

	Internal diameter of pipe	Thickness of body	Thickness of boss	Length of boss	Thickness of flange finished	Thickness of flange rough	Diameter of bolt holes	Outside diameter of flange	Diameter of bolt circle	Number of bolts	Diameter of bolts
For 75 lbs. pressure or 170 feet of head	3	·328	·40	1·25	·50	·56	·55	6½	5¼	4	1/2
	3½	·341	·42	1·28	·51	·57	·61	7¼	5 9/16	4	9/16
	4	·354	·43	1·30	·53	·59	·61	8	6 7/16	5	9/16
	5	·380	·46	1·35	·56	·63	·61	9	7 3/8	6	9/16
	6	·406	·49	1·40	·60	·67	·68	10¼	8 11/16	6	5/8
	8	·458	·55	1·50	·66	·74	·68	12½	10 9/16	8	5/8
	10	·510	·61	1·60	·73	·81	·81	15	13 3/16	10	5/8
	12	·563	·67	1·70	·80	·89	·93	17¾	15 9/16	10	3/4
	16	·667	·79	1·90	·93	1·0	·93	22	19 5/16	14	7/8
For 100 lbs. pressure or 230 feet of head	3	·383	·55	1·25	·72	·80	·61	7½	6	4	9/16
	3½	·398	·56	1·28	·74	·82	·61	8	6½	5	9/16
	4	·414	·58	1·30	·76	·84	·68	9	7¼	5	5/8
	5	·444	·62	1·35	·80	·89	·68	10	8 9/16	6	5/8
	6	·474	·65	1·40	·84	·93	·68	11	9 13/16	6	5/8
	8	·535	·72	1·50	·92	1·02	·68	13½	11¾	8	5/8
	10	·596	·79	1·60	1·00	1·11	·81	16	14	10	3/4
	12	·657	·86	1·70	1·08	1·20	·93	19	16¾	10	7/8
	16	·778	1·00	1·90	1·24	1·38	·93	23½	21	14	7/8

Fig. 6 shows the ordinary forms and denominations of pipe bends.

15. Special forms of pipe joint.—Fig. 7 shows an adjustable form of pipe joint. When the angle between two pipes

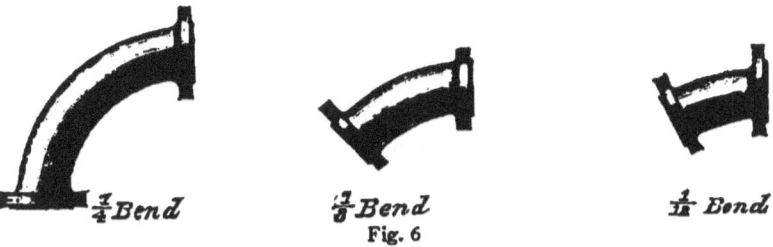

Fig. 6

varies or is not definitely known beforehand a joint of this kind, which can be arranged at any angle between 80° and 180°, is convenient.

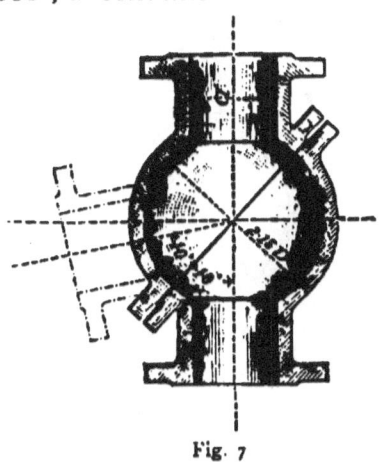

Fig. 7

Fig. 8

Fig. 8 shows a pipe joint with one flange loose. The joint is made tight by a lead ring.

Fig. 9 shows what is termed a lens joint, which is easily made tight and which adjusts itself to small changes of direction of the pipe. The joint is made by a gunmetal ring with spherical surfaces.

Fig. 9

16. Joints for hydraulic mains.—Fig. 10 shows the joint

used by Sir W. Armstrong for the pipes of his accumulator. These pipes are subjected to the enormous water-pressure of 700 lbs. per sq. in. The pipes are of the best remelted cast iron, and are tested to 2,500 lbs. per sq. in. When 5 ins. diameter they are $\tfrac{7}{8}$ in. thick. Each end of the pipe has two strong elliptical flanges, with two bolts. One pipe slightly enters into the other, forming a dovetailed recess, in which is placed a gutta-percha ring, $\tfrac{1}{4}$ in. thick. The thickness of these pipes may be calculated by the rule—

$$t = \cdot 000178 dp + \tfrac{1}{4},$$

the working stress in the cast iron being about 2,800 lbs.

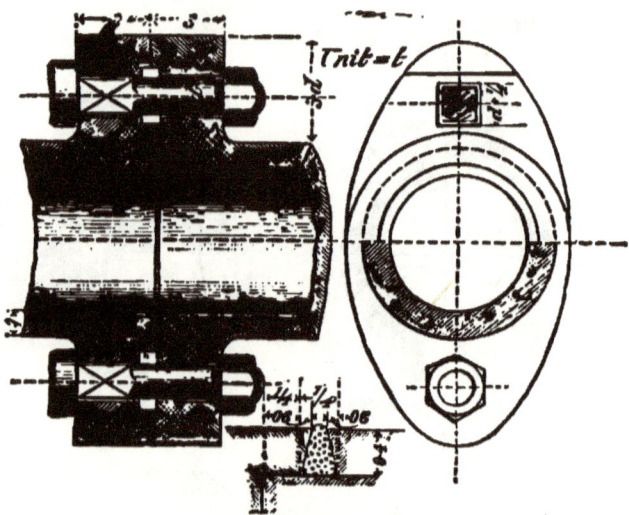

Fig. 10

per sq. in., and $\tfrac{1}{4}$ in. being allowed for inequalities of casting and corrosion. The bolts are of such a diameter that the stress at reduced section of thread is 7,700 lbs. per sq. in.

17. *Cylinders for great internal pressure.*—In the construction of vessels for great hydraulic or other pressure special difficulties arise. If made of boiler plate, the riveted joints diminish the strength of the vessel and limit the thickness

of the plates which can be used. If cast iron is adopted, the vessel must be ponderous in consequence of the low tenacity of the material, and the fluid sometimes escapes through porous parts of the casting. Dr. Siemens has described the construction of an air reservoir, to sustain 1,000 lbs. pressure per sq. in., of steel rings. These rings (40 in. in diameter and 12 in. in depth) were rolled out of ingots in a tyre mill, and had a slight flange (fig. 11) at their edges. The ends of the reservoir were made hemispherical, and beaten out of steel plate. The joints were made by turning a V-groove in the faces of the rings and placing in it a packing ring made of $\tfrac{5}{10}$ annealed copper wire. The rings and ends were held together by 20 steel longitudinal bolts, passing through two rings bearing on the hemispherical ends of the cylinder. These bolts were $1\tfrac{1}{4}$ in. in diameter with enlarged screwed ends. The rings were of steel, having a tenacity of 45 tons per sq. in., and were loaded to a working stress of 15 tons per sq. in.

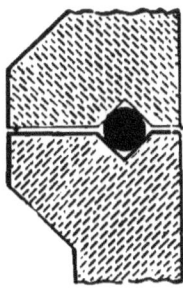

Fig. 11

18. *Socket joints.*—Socket pipes should be cast vertically with the socket at bottom. Socket pipes are jointed either

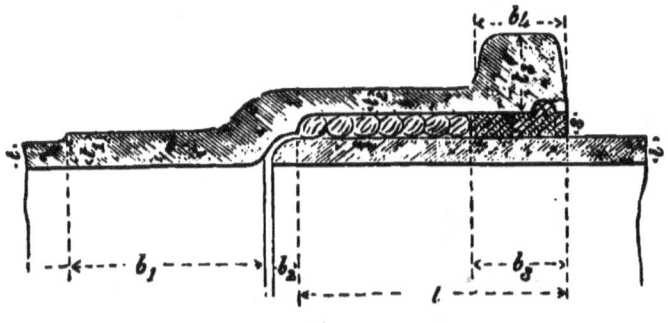

Fig. 12

with a gasket and lead joint, a rust joint, or a bored and turned joint. Fig. 12 shows an ordinary lead joint. When the pipes are in place, a few coils of gasket or tarred rope are driven into the socket. Clay is then put round the outside of the socket and a lead ring is cast in it. The clay is removed and the lead stemmed tightly into the socket. The proportions may be as follows : Let $t=$ thickness, and $d=$ diameter of pipe ;

$$t_1 = 1\cdot07\ t + \tfrac{1}{16}$$
$$t_2 = 0\cdot025\ d + \tfrac{1}{4} \text{ to } 0\cdot025\ d + 0\cdot6$$
$$t_3 = 0\cdot045\ d + 0\cdot8$$
$$s = 0\cdot01\ d + \cdot25 \text{ to } 0\cdot01\ d + \cdot375$$
$$b_1 = 0\cdot075\ d + 2\tfrac{1}{4}$$
$$b_2 = t_2$$
$$l = 0\cdot09\ d + 2\tfrac{3}{4} \text{ to } 0\cdot1\ d + 3$$
$$b_4 = b_3 = 0\cdot03\ d + 1$$

A rust joint is very similar to a lead joint except that iron cement is stemmed in with a cold chisel in place of the lead.

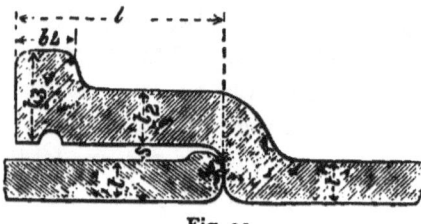

Fig. 13

The iron cement consists of cast-iron borings or turnings, which should be passed through a sieve of eight meshes to the inch. One ounce of sal-ammoniac is added to each hundredweight of cast iron, and the mass is damped. When it has heated, it may be kept for some time in water.

Fig. 13 shows a different form of socket. The proportions may be the same as those just given.

Fig. 14

Fig. 15

Figs. 14 and 15 show two forms of bored and turned socket

and spigot joints. When the bored and turned part is long the pipes are rigid, and are liable to be broken by the earth pressure. Hence, the fitting part is now often only ⅝ in. in length, and has a very slight taper. The joint is made by painting over the faced part with red lead, or with fresh and liquid Portland cement. The pipe is then put in place, and driven home by a wooden mallet, or by swinging the next length of pipe. The socket is filled up with cement. Joints of this kind are more easily and quickly made than lead joints, but lead joints are preferable in passing round curves. Socket pipes should be cast with the socket downwards, and about a foot of length should be allowed at the spigot end, into which the scoriæ may rise, and which is broken off when the pipe is cast.

19. With socket pipes care must be taken to provide against the end thrust due to the water pressure at bends. At a right-angled bend, the total pressure in the direction of the tangents to the bend on a pipe of diameter d feet, through which water is passing under a pressure head of h feet, and with a velocity of v feet per second, is

$$\frac{\pi}{4} d^2 G \left(h + \frac{v^2}{g} \right) \text{lbs.}$$

Where $G = 62 \cdot 4$ lbs. Hence the resultant pressure bisecting the angle of the bend is

$$1 \cdot 414 \frac{\pi}{4} d^2 G \left(h + \frac{v^2}{g} \right) \text{lbs.} = 69 \cdot 29 \, d^2 \left(h + \frac{v^2}{g} \right).$$

This must be provided against by casting a foot on the pipe and abutting it against a block of masonry.

A good joint for pipes used for temporary purposes is shown in fig. 16. A loose tapered double socket is fitted over the ends of the pipes. The joints are made by indiarubber rings driven into the sockets.

20. *Joints for lead pipes.*—Lead pipes are useful, because they are easily bent. They are manufactured by drawing them over a mandril by hydraulic pressure. They are

sometimes lined with tin, when used to convey water which dissolves the lead. Joints may be made by flanging out the ends of the pipes, and compressing these flanges between two iron rings, with bolts. Commonly, the joint is made by soldering, and is termed a 'plumber's wiped joint' (fig. 17).

21. Fig. 18 shows two ways of making a socket joint for wrought-iron pipes. The sockets are of cast iron.

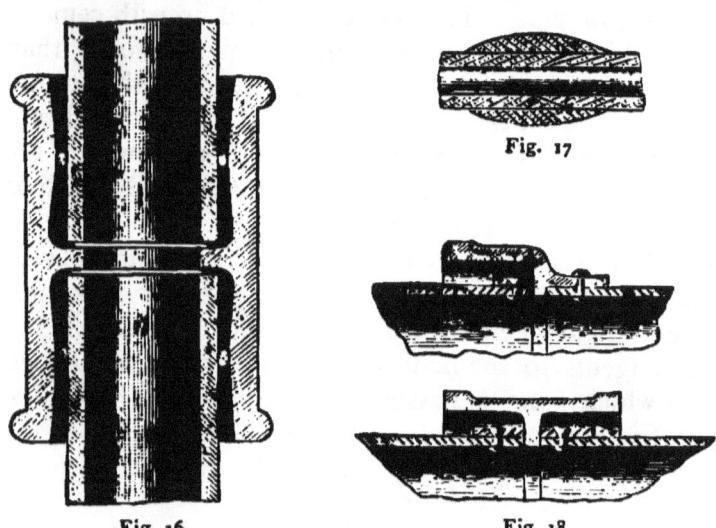

Fig. 17

Fig. 16   Fig. 18

Wrought iron is chiefly used for very large or very small pipes. From its thinness, it is liable to more injury from

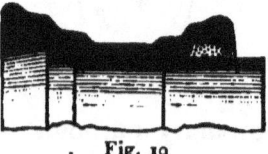

Fig. 19

corrosion than cast iron, and hence it has been suggested that large wrought-iron mains should be lined with a thin coating of Portland cement.

Fig. 19 shows another form of ring joint for lead used with wrought-iron pipes. Special rings and sockets are now made also of thin rolled steel.

Joints for wrought-iron and steel tubes.—Fig. 20 shows the ordinary forms of joints. At *a* one pipe is bulged and the other screwed into it. At *b* an internal screwed ferrule or socket piece is inserted. At *c*, an external socket is shown screwed on both pipe ends, and at *d* a similar socket piece of lighter form.

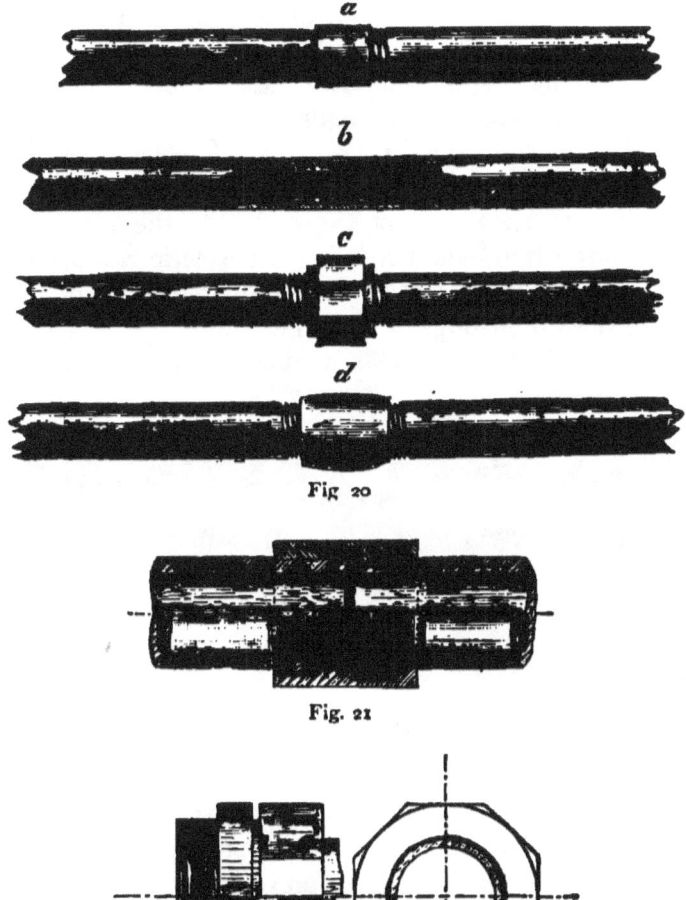

Fig. 20

Fig. 21

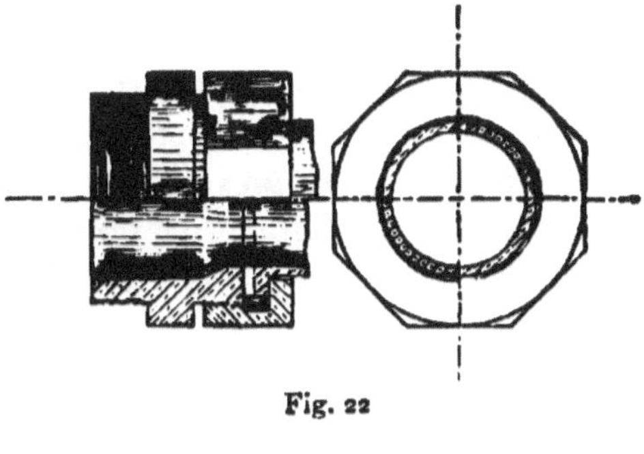

Fig. 22

A very good form of joint is shown in fig. 21, where security against leakage under great pressure is necessary.

The two pipe ends are screwed with right- and left-handed threads. The end of one pipe is turned with a flat face and the other with a sharp V-shaped edge. The socket or coupling piece then draws the ends together into metallic contact. A union joint, shown in fig. 22, may also be used, with a packing ring of leather or india-rubber, if necessary.

22. *Boiler tubes.*—The modes of fixing boiler tubes in tube plates is shown in fig. 23. The lower figure shows an ordinary boiler tube, fixed at one end by slightly enlarging the tube and riveting over the end. At the other end a wrought-iron ferrule is driven in. In American locomotives cast-iron ferrules have been used and are said to be tighter than ferrules of wrought iron. The upper figure is a boiler tube adapted also to act as a longitudinal stay rod. For

STAY TUBE WITH BACKNUTS IN BOILER

BOILER TUBE WITH FERRULE DRIVEN IN END

Fig. 23

this purpose the ends are screwed and the tube fixed by a pair of thin nuts at each end. Sometimes one end of the stay tube is enlarged (fig. 1). This facilitates getting it into place.

Boiler and stay tubes are subjected to external pressure and must be calculated by the rules for resistance to collapse. But practical conditions prevent any close adherence to theoretical proportions. Ordinarily boiler tubes are $2\frac{1}{2}$ to 4 inches in diameter and their thickness for steam pressures above 40 lbs. per sq. in. is

$$t = \frac{6,000}{pd} + \frac{1}{10}$$

Stay tubes are very commonly $\frac{1}{4}$ in. thick to allow for cutting the screw-threads at the ends.

23. *Condenser tubes.*—Condenser tubes are now generally made of brass. Sometimes they are tinned as a protection against the action of fatty acids. They are from ½ in. in diameter in small condensers to 1 in. in very large condensers and about 0·05 in. thick. They are fixed in the tube plate in different ways. Sometimes a simple soft wood ferrule is driven between tube and tube plate. The ends of the tubes are split and expanded to prevent creeping.

Fig. 24

Generally some very simple form of stuffing-box is used, which can be easily and cheaply made. Perhaps the simplest method is that shown in fig. 24. The tube plate is drilled with recesses round the tube ends which form stuffing-boxes. The gland is simply a short piece of screwed tube cut off by a saw. The packing is a ring of tape. This method permits the free expansion of the condenser tube to occur without causing leakage.

CHAPTER II

ARRANGEMENT AND PROPORTIONS OF STEAM-ENGINE CYLINDERS

24. It is beyond the scope of this treatise to consider steam engines as machines for transforming heat into mechanical energy and the thermo-dynamic laws of the action of steam. Nevertheless, it is impossible to deal rationally with the details of steam engines unless the mechanical action of the steam in the cylinder is understood. The principal straining actions on which the dimensions of engine parts depend are due initially to the steam pressure on the piston, and the designer must be able to determine those straining actions, at least approximately, for any position of the mechanism. For a single-cylinder engine the problem is comparatively simple; but if a multiple-cylinder engine is used, then the problem is more difficult. Further, various problems in designing arise out of the action of the inertia of the moving parts, and cannot be attacked without a general knowledge of the arrangement of the mechanism of the engine. Hence it seems desirable to discuss from a standpoint as purely mechanical as possible the general arrangement and types of steam engines, as a basis for proportioning engine details.

25. *Arrangement of mechanism.*—(*a*) Beam engines, a type still retained in large pumping engines. In marine engines the beam engine took the form known as the side-lever engine, a type which has disappeared. (*b*) Ordinary direct-acting engines, with a connecting rod between the crosshead and crank. (*c*) Trunk engines, in which the connecting rod is attached direct to the piston inside a

trunk which forms the sliding guide. (*d*) Oscillating engines which kinematically are merely an inversion of the direct-acting engine, the connecting rod of the latter becoming the frame link of the former. (*e*) Return connecting-rod engines are a variety of type (*b*) in which space is saved by bringing the crank shaft close to the cylinder and placing the slide and cylinder on opposite sides of the crank shaft.

If s is the stroke and d the diameter of the cylinder, the length over all from centre of crank shaft to back of cylinder may be reduced to $3.5s + 0.6d$ in direct-acting engines; $2.6s + 0.6d$ in trunk engines; and $1.6s + 0.8d$ in return connecting-rod engines.

Combined and compound engine arrangements.—With simple engines it is least costly to use a single engine, and such an engine has the fewest working parts—a not inconsiderable advantage. But for various reasons a combined engine is often adopted. With two engines acting on cranks at right angles, and, still better, with three engines acting on cranks at 120°, the turning moment is much more uniform than with a single engine; the cylinders are smaller in diameter and the working parts lighter. With a single engine there are two dead points in the revolution, and consequently positions in which the engine will not start by the action of the steam pressure. Pairs of simple engines may be placed parallel or inclined. The latter arrangement was at one time adopted because solid crank shafts with cranks at right angles were difficult to obtain, and the inclined or diagonal arrangement permitted both engines to act on a single overhung crank-pin, often on a crank arm keyed to the shaft.

Compound engines, or engines in which the steam expands successively in two cylinders, are of two general types: (1) the Woolf engine (or, more strictly, Roentgen engine), and (2) the Receiver engine. In the Woolf engine (*a*, fig. 25) the steam passes as directly as possible from one end of the high-pressure to one end of the low-pressure cylinder.

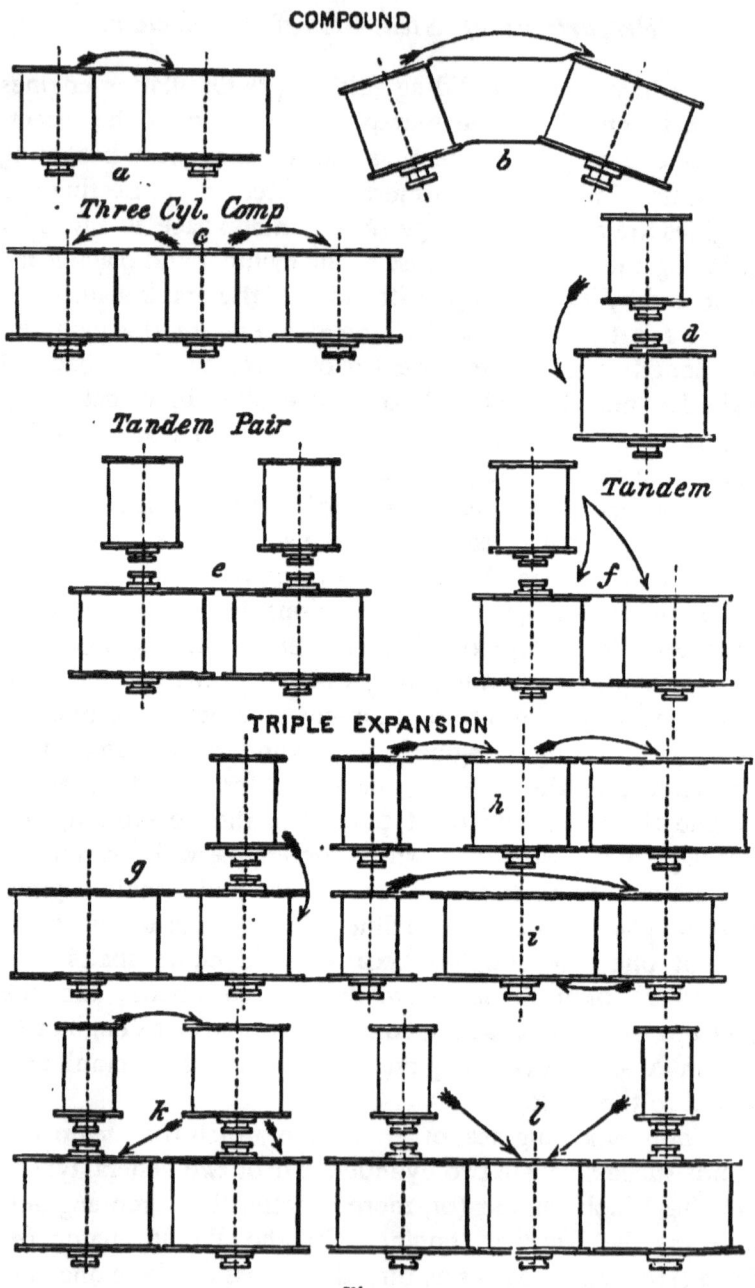

Fig. 25

In that case the cranks are usually at 180°. Some designers have attempted to reduce the action of the unavoidable passage between the cylinders by placing the cranks at 160° and allowing the HP cylinder to begin exhausting a little before the LP opens to steam. If the cylinders are placed tandem, as at *d*, there is only a single crank on which both pistons act. To get greater uniformity of twisting moment with a Woolf engine, a pair of tandem cylinders may be used, as at *e*. If in the arrangement *a* the cranks are at right angles, then there must be an intermediate reservoir into which the steam can expand till the LP cylinder opens. This is the simplest form of receiver engine. The arrangement *b*, with inclined cylinders at right angles, both pistons acting on a single crank, is an equivalent arrangement. In some cases it is convenient to divide the low-pressure cylinder into two cylinders, and then the arrangement *c* or *f* may be used.

The prejudicial action of the cylinder walls, causing condensation, increases rapidly with the range of temperature to which they are exposed. Hence, engineers have been driven, when using high pressures and high ratios of expansion, to carry the principle of compounding further, and to expand the steam in stages in three or even four successive cylinders. The simpler arrangements of triple expansion engines are shown in fig. 25, at *g*, *h*, *i*, *k*, *l*.

26. *Working steam pressures.*—Of course no definite rule can be given, but the following are about the usual working boiler pressures used now in the most economical engines of given types:—

Type of Engine	Steam Pressure in lbs. per sq. in.	
	Non-condensing	Condensing
Simple	85	60
Compound	120	90
Triple	180	150
Locomotives, simple	150	—
,, compound	180	—

The initial pressure in the cylinder will be less than the boiler pressure by the amount of resistance in the steam pipe and passages. The loss of pressure may be taken to be 7 to 10 per cent. of the boiler pressure.

27. *Piston speed*.—The piston speed is determined partly from considerations of local convenience, and partly from consideration of the work to be done and the durability desired. The piston speed increases a little, other things being equal, with the length of stroke. Let $s=$ length of stroke in feet, $v=$ piston speed in feet per minute. Then

$$v = a\sqrt[3]{s}$$

where a has the following values in different cases.

Type of engine	$a =$
Direct-acting pumping engines (without fly wheel)	80 to 120
Beam pumping engines	90 to 200
Horizontal Corliss engines	220 to 400
Horizontal compound mill engines	200 to 400
Small horizontal engines, ordinary	240 to 300
Short stroke, quick speed	400 to 550
Locomotive engines (highest speed)	800 to 1,000
Quick speed, short stroke, single-acting engines	550 to 650
Paddle marine engines	206
Screw marine engines	330

These figures give speeds of about 125 feet per minute in slow pumping engines; 240 to 360 in beam pumping engines; 300 to 450 in ordinary horizontal engines; 500 in single-acting quick-speed engines; 700 in marine screw engines; 1,000 in locomotives; that is, taking cases such as most ordinarily occur.

28. *Ratios of cylinder volumes in compound engines*.— The cylinder volume is understood to mean the product of the area of piston and length of stroke or volume described by the piston in a single stroke. In proportioning the relative volumes of compound engine cylinders, designers have generally aimed at making either the range of temperature in each cylinder equal, or the effective work done in each cylinder equal. No very simple rule can be given to secure either result, because the clearance and receiver spaces affect the action of the steam so considerably. A means of testing any given arrangement will be described presently, meanwhile the following table of proportions usual in existing engines will be of service :—

ORDINARY RATIOS OF CYLINDERS IN COMPOUND AND MULTIPLE EXPANSION ENGINES

	Ratio of Cylinder Volumes			
Compound				$\frac{HP}{LP}$ 3 to 4
Triple		$\frac{HP}{IP}$ 2 to 2·75	$\frac{IP}{LP}$ 2 to 4	$\frac{HP}{LP}$ 5 to 8
Quadruple	$\frac{1st}{2nd}$ 1·5 to 2	$\frac{2nd}{3rd}$ 1·7 to 2¼	$\frac{3rd}{4th}$ 2 to 3	$\frac{1st}{4th}$ 6 to 12

29. *Construction and proportions of steam-engine cylinders.*—A simple form of cylinder is shown in fig. 26. The valve chest is arranged for a slide valve, the steam and exhaust passages are cast in one with the barrel. The front end of cylinder and valve chest are closed by covers.

The thickness of the cylinder barrel must be determined so that it is (*a*) strong enough to resist the internal steam pressure; (*b*) rigid enough to prevent any sensible alteration of form; (*c*) it must be thick enough to insure a sound casting; and (*d*) thick enough to permit reboring once or twice when worn. A rule has been already given (§ 6), which makes the thickness depend on the steam pressure.

Generally other considerations than strength are of so much importance, that the following empirical rule agrees better with practice :—

Thickness of barrel of cylinder

$$= t = 0 \cdot 02 d + 0 \cdot 5 \text{ to } 0 \cdot 02 d + 0 \cdot 75.$$

The flanges of the cylinder have a thickness $1 \cdot 3 t$; the metal of the valve chest and passages $0 \cdot 7 t$; valve chest flanges t; cylinder face in valve chest $1 \cdot 25 t$. Sometimes a false face is used (fig. 28), screwed to the cylinder face with gunmetal screws. Then the cylinder face may have a thickness t, and the false face a thickness $0 \cdot 8 t$ if of cast iron, and $0 \cdot 6 t$ if of gunmetal.

Many engine cylinders have a steam jacket round the cylinder barrel, and this is either cast in one with the barrel (fig. 27), or formed by a liner inserted steam-tight in the barrel (fig. 28). When cast in one with the barrel, the jacket is often omitted from the space below the steam passages, where it is shown in fig. 27. There is less danger of leakage if the jacket is cast with the cylinder barrel, but, of course, the moulding of the cylinder is more difficult. If the jacket is made by a liner, the liner may be of cast iron or steel. A cast-iron liner may have a thickness $0 \cdot 8 t$, but a steel liner may have a thickness $pd/3000$. The liner may be fixed as shown in fig. 28, with a flange and studs at the front end of cylinder. At the back end it is turned to fit the barrel tightly, and a recess is formed packed with asbestos. Fig. 29 shows a liner held in place by the pressure of the cylinder cover.

The front end of the cylinder should be rigidly bolted to the engine framing. At the back end the bolt-holes should be enlarged so that the cylinder can expand and contract. The expansion will not exceed 1-500th of the distance between the front and back bolts. The bolts may be calculated to resist the maximum pressure on the piston with a stress not exceeding 3,000 to 4,000 lbs. per sq in.

Proportions of Steam-Engine Cylinders 35

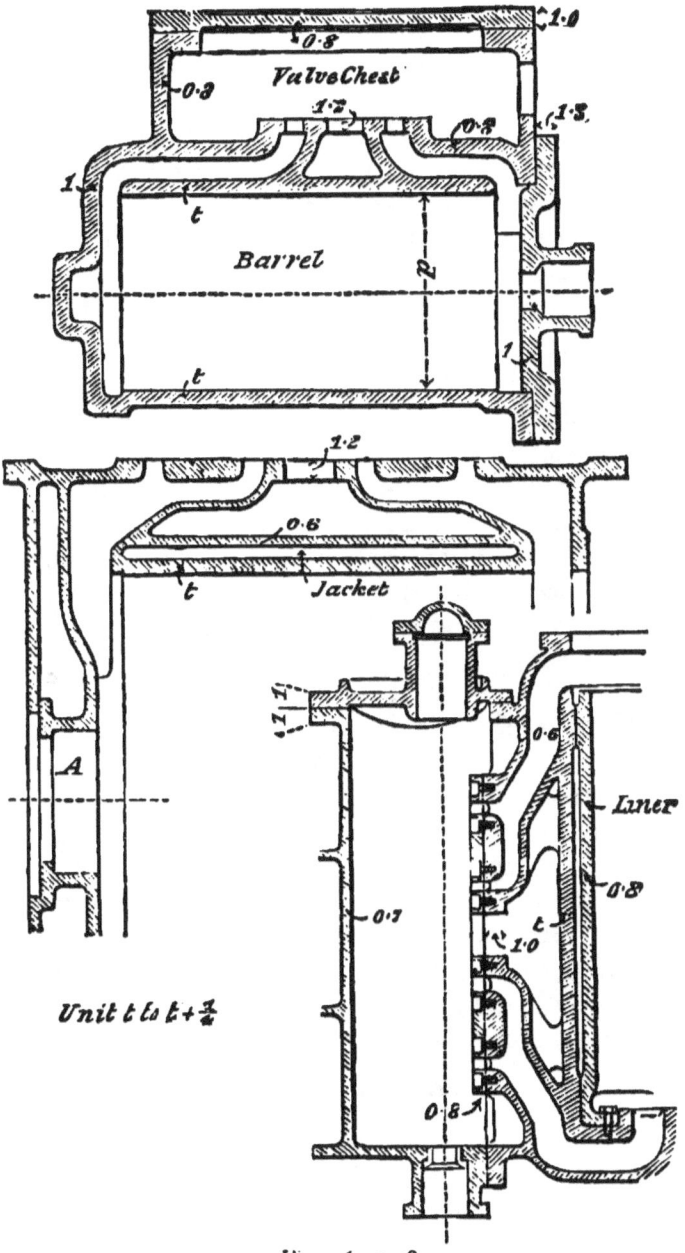

Figs. 26, 27, 28

When the cylinder covers are not very large, they have a single thickness stiffened outside by radiating ribs. Larger covers are made hollow with a double thickness of metal and stiffening ribs between. The cylinder cover bolts or studs may be calculated for a strain of 4,000 lbs. per sq. in. of net section, which allows a margin to resist stresses due to priming. (See I. § 82.) The back end of large cylinders is cast with an aperture (A, fig. 27), to admit the boring bar, and this is closed by a cover. The pitch of studs and bolts may be about $4\sqrt{t_1}$ to $5\sqrt{t_1}$ where t_1 is the thickness of metal in the cover.

Fig. 29 shows a small engine cylinder with the arrangement of the various fittings.

30. *Clearance.*—Clearance distance must be left between the piston at the end of the stroke and the cylinder covers as a security against the displacement of the piston by the wear of its connections. The distance in inches left between the piston and cylinder cover at each end of the stroke may vary from $\frac{1}{8} + \frac{n}{15}$ in small engines to $\frac{1}{4} + \frac{n}{10}$ in large engines, where n is the number of joints subject to wear between piston and crank shaft.

The clearance space has an important influence on the action of the steam, and as the steam passages form part of the clearance space (regarded as space occupied by steam during expansion), they are reckoned with the cylinder clearance in estimating the clearance volume, or, as it is usually termed, simply, the clearance. The clearance volume is most conveniently given as a fraction or percentage of the volume described by the piston in a single stroke.

Clearance Volume

Simple engines	Percentage of Cylinder volume
Corliss valves	2 to 4
Double-beat valves	5 to 7
Long slide with short passages	1·8 to 2·7
Ordinary slide	6 to 12

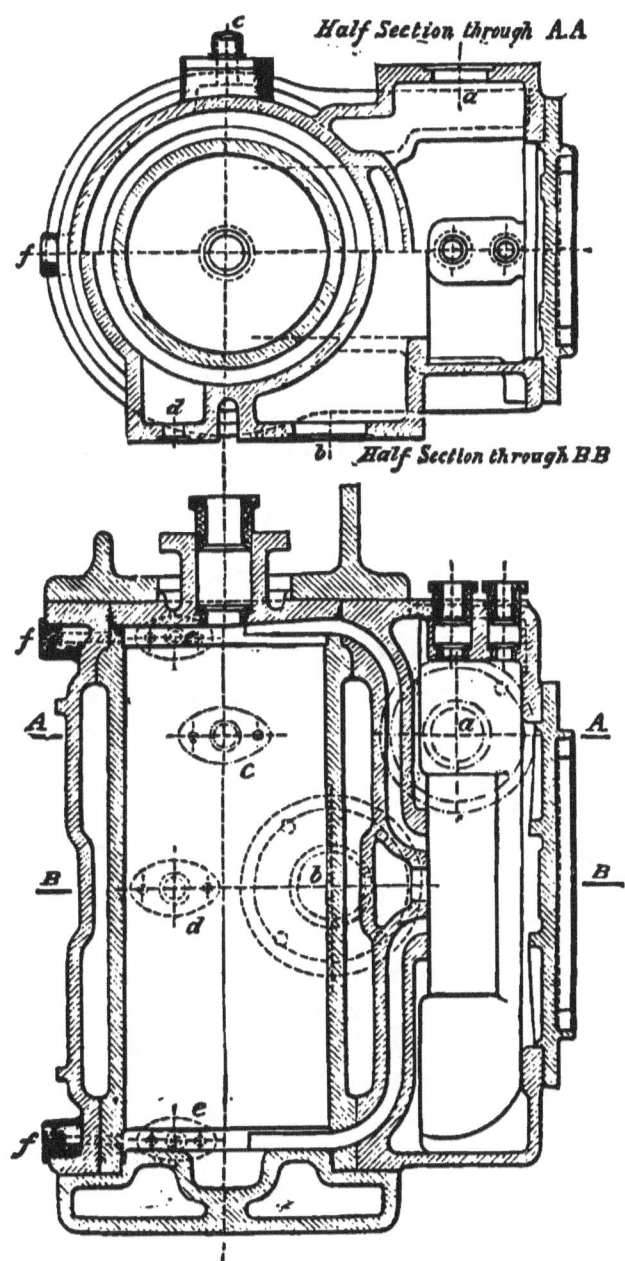

Fig. 29

a, Steam pipe ; *b*, Exhaust ; *c*, Jacket steam pipe ; *d*, Jacket drain ; *e*, Cylinder drain ; *f*, Indicator.

Compound	HP		LP	
Woolf with long slide and short passages	3 to 4		2 to 3	
Ordinary slides	7 to 12		5 to 10	
Gridiron slides	5		3	

Triple expansion	HP	IP	LP
Slide valves	10 to $12\tfrac{1}{2}$	5 to 10	5 to 7

31. Hydraulic press cylinders.—When the thickness of the cylinder is not small compared with the radius, the following equation must be used (Part I. p. 48):—

$$t = \frac{d}{2}\left\{-1 + \sqrt{\frac{3f+2p}{3f-4p}}\right\}$$

where t is the thickness of metal in the cylinder; d its internal diameter; f the safe working stress on the material, and p the excess of internal over external pressure.

Steel	$f =$	15,000
Wrought iron		12,000
Cast iron		5,000

In many cases p is fixed by the conditions of the case. Then if the total effort of the ram P is also fixed,

$$\frac{d}{2} = \sqrt{\frac{P}{p\pi}},$$

and then t is ascertained by the equation above. Hermann has shown that there is a value of p for which the external diameter of the press cylinder is least, and obviously this makes the construction of the press cheap. Let D be the external diameter of the press. Then

$$D = d + 2t = d\sqrt{\frac{3f+2p}{3f-4p}}$$

Let $x = p/f$.

$$\frac{D^2}{d^2} = \frac{3f+2p}{3f-4p} = \frac{3+2x}{3-4x}$$

and putting $d^2 = 4P/\pi p$.

$$\frac{D}{2} = \sqrt{} \sqrt{\frac{P}{\pi p}} \sqrt{\frac{3+2x}{3-4x}}$$

$$= \sqrt{} \sqrt{\frac{P}{\pi f}} \sqrt{\frac{3+2x}{x(3-4x)}}.$$

This will be a minimum for $x = p/f = 0\cdot336$. In other words, with the above values of the working stress, the working pressure should be 5,040 lbs. per sq. in. for steel; 4,030 for wrought iron, and 1,700 for cast iron. In any case, if $x = 0\cdot336$,

$$D = 1\cdot5d. \text{ nearly}.$$

32. *Indicator diagram. Single cylinder engine.*—By using an indicator on any actual engine a curve is drawn, the co-ordinates of which are the pressure and volume of steam in the cylinder at each point of the stroke. For a complete revolution the curve is a closed curve, the area of which is proportional to the work done by the steam on one side of the piston. In double-acting engines a corresponding diagram gives the work done on the other side of the piston. The mean vertical width of the diagram is called the mean effective steam pressure. It will be denoted by p_m.

If p_m is the mean effective pressure in lbs. per sq. in., A, the area of the piston in sq. ins., then p_m A lbs. is the mean effort driving the piston. If N is the number of revolutions per minute, l the length of stroke in feet, $2 p_m$ A N l foot lbs. is the work done per minute, and the indicated horse power is—

$$\text{I. H. P.} = \frac{2 p_m \text{ A N } l}{33000}.$$

Now, in designing an engine it is frequently convenient to draw an ideal indicator diagram, either in order to get a value for the mean effective pressure, or to determine the forces acting at any given point of the stroke. In drawing such a diagram, so long as mechanical and not thermo-dynamical problems are in question, it is almost always accurate enough to assume that the expansion curve is an hyperbola.

To draw an ideal diagram, take $oa = cl$ (fig. 30), the length of stroke equivalent to the clearance; ab the length of stroke, to any convenient scale of lengths. Now take ac on any other convenient scale to represent the initial absolute steam pressure (that is, gauge pressure + 14.7 lbs. per sq. in.). The initial pressure may be assumed at 7 to 10 per cent. less than the boiler pressure, if there is no special

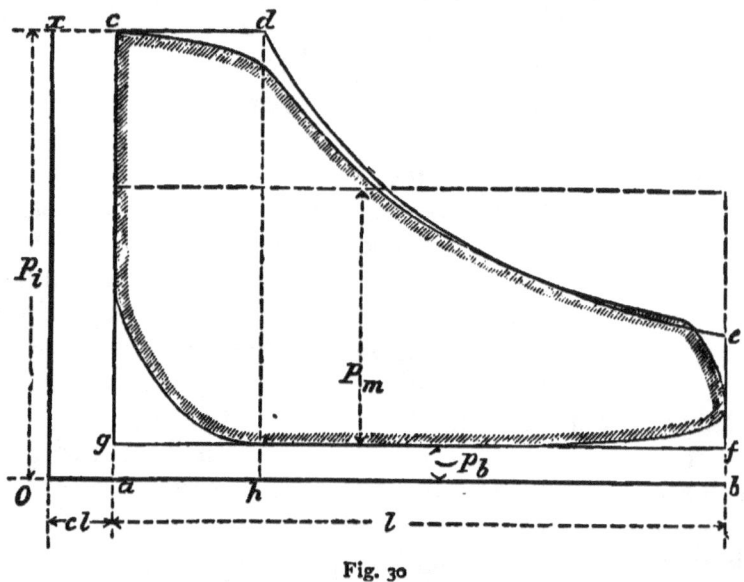

Fig. 30

cause of loss. Now let cd be the portion of the stroke, during which steam is admitted, so that ab/cd is the ratio of expansion. From d draw an hyperbolic expansion curve de, having ox and ob as asymptotes. For any point of the curve the product of the co-ordinates $= oh \times hd$. Set up ag, the back pressure p_b, which may be taken at 2 to $2\frac{1}{2}$ lbs. per sq. in. for condensing engines and 16 or $16\frac{1}{2}$ for non-condensing engines. Then $cdefg$ is the theoretical indicator-diagram for the conditions assumed. The actual diagram will have some such form as is shown by the shaded figure, and its area will be less than that of the theoretical diagram in a ratio to be determined presently. This ratio will be called the diagram factor ι. The nominal ratio of expansion is $ab/cd = r$. But the real ratio of expansion is ob/xd, and, in consequence of the clearance, this is less than the nominal ratio, sometimes considerably less. If p_m is the mean vertical breadth of the diagram, the work done per single stroke is the rectangle $p_m \times gf$, and it is often convenient to deal with the mean pressure p_m instead of the diagram area, because we have not then to consider the unit of area of the diagram, which represents a unit of work.

The value of p_m can be obtained directly by calculation. The real ratio of expansion is—

$$\frac{l + cl}{\frac{l}{r} + cl} = \frac{1 + c}{\frac{1}{r} + c}$$

From the known properties of the hyperbola, the area $oxdeb$ is—

$$p_1 \left(\frac{1}{r} + c\right) l \left(1 + \log r\right)$$

where the logarithm is hyperbolic; or

$$p_1 \left(\frac{1}{r} + c\right) l \left(1 + 2\cdot3 \log r\right)$$

where the logarithm is a common logarithm. Hence the area $gcdef$ is

$$p_1 \left(\frac{1}{r} + c\right) l \left(1 + 2\cdot3 \log r\right) - p_1 cl - p_b l.$$

The mean breadth of the diagram, or mean effective pressure, is

$$p_m = p_1 \left\{\left(\frac{1}{r} + c\right) \left(1 + 2\cdot3 \log r\right) - c\right\} - p_b.$$

The diagram-factor for single-cylinder engines varies

from 0·8 to 0·9, so that the actual mean effective pressure to be expected in an engine will be 0·8 to 0·9 of the value given by the formula above. Generally it may be taken at 0·85 of the value above for slide-valve engines of moderate size and 0·9 for large engines. If there is specially large compression or wire-drawing the diagram-factor may fall to 0·7 or lower.

Having now the means of determining the probable mean pressure for any assigned conditions of working, the I. H. P. of a given engine, or the area of piston for a given I. H. P. can be determined by the formula given above.

It is desirable to point out that the vertical breadth of the diagram (fig. 30) is the pressure at a given moment of the forward stroke, less the pressure on the same side of the piston at the corresponding point of the return stroke. In those cases in which we have to consider the effort transmitted from the piston to the crank pin we require to know the pressure on one side of the piston, less the pressure at the same moment on the other side. To get this the back-pressure line must be drawn reversed. It is the same thing if the forward-pressure line of one end of the cylinder is combined with the back-pressure line of the other end.

33. *Provisional determination of mean effective pressure in a compound engine.*—In the compound-engine diagram it is convenient that the horizontal abscissæ should represent volumes of steam in the cylinder. Take ox (fig. 31) to represent the initial steam pressure to any scale, oa the clearance volume cv in the low-pressure cylinder, ab the volume described by low-pressure piston. Let cd be the volume described by high-pressure piston at cut-off, so that ab/cd is the intended nominal ratio of expansion. If the diagram is drawn for the purpose of determining the cylinder volumes, any distances, ab, cd, may be taken provided they are in the given ratio. Now draw the hyperbola dge and the back-pressure line lm. Then $cdeml$ will be the total

work of the engine, and its mean vertical breadth will be the mean effective pressure *reduced to the area of the low-pressure piston.* From the mean pressure so obtained 10 to 20 per cent. should be deducted to allow for the effects of compression, wire-drawing and waste expansion;

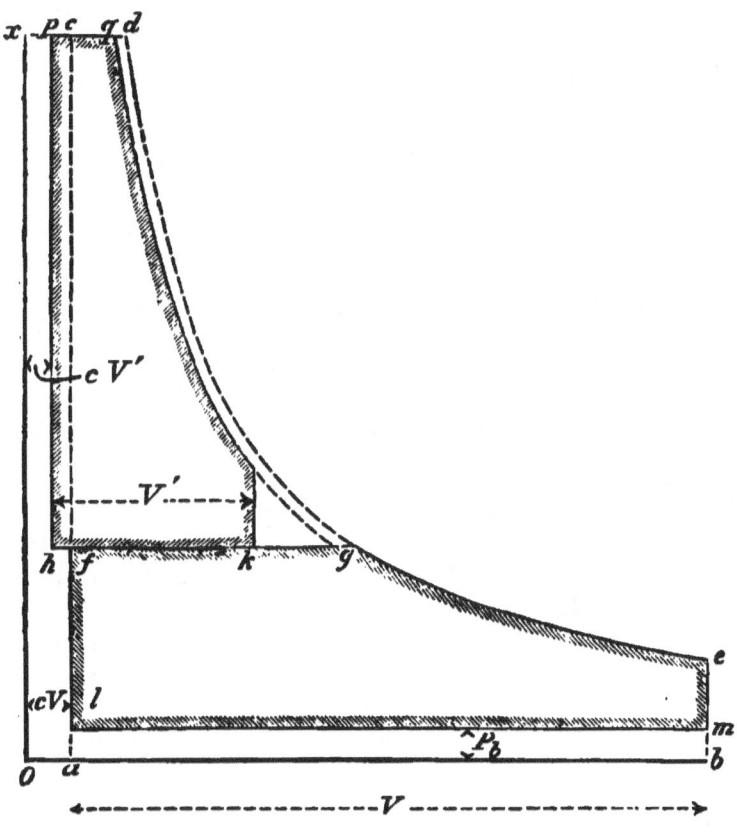

Fig. 31

or, what is the same thing, the area should be multiplied by a diagram-factor ϵ, such as is given below. The area divided by lm is the mean effective pressure *reduced to the low-pressure piston.*

Let p_m be this mean effective pressure, corrected as

indicated, and A the area of the low-pressure piston. Then from the relation

$$\text{I. H. P.} = \frac{2 p_m \text{ A N } l}{33000}$$

similar to that for simple engines given above, we can determine the I. H. P. if the diameter of low-pressure cylinder is known, or the diameter if the I. H. P. is given.

If the clearance of the high-pressure cylinder is different from that of the low-pressure cylinder, set off $xp = c'v'$, the high-pressure clearance volume, and take $pq = cd$, the admission volume to high-pressure cylinder. An hyperbola through q is the high-pressure expansion curve. Take hk/pq, the required ratio of expansion in high-pressure cylinder. Then $pqkh$ is the high-pressure diagram, which does not much differ in area from $cdgf$. The line hg is usually drawn so as to approximately equalise the areas of the high- and the low-pressure diagrams. From the diagram now completed the mean effective pressures in each cylinder can now be obtained, and these may be corrected for compression and wire-drawing as before.

The diagram-factor ϵ, or ratio of the area of a pair of actual diagrams to the area $cdeml$, has been found for certain cases to have the following values:—

	Diagram Factor $\epsilon =$
Compound-receiver horizontal engine, with slide-valve, jacketed . .	0·93
,, ,, ,, unjacketed . .	0·83
Tandem Woolf engines . . .	0·75 to 0·83
Worthington pumping-engine cylinders and receiver, jacketed . .	0·97

The total nominal ratio of expansion in the diagram above is ab/pq and the real total ratio of expansion is ob/xq very nearly. If, as is usual, the two pistons have the same stroke, then hk and ab are proportional to the areas of the high-pressure and low-pressure pistons. A similar process will serve in provisionally designing a triple-expansion engine.

34. *More accurate construction of the indicator diagrams for a compound engine. Case I. Woolf engine.*—Supposing the sizes of cylinders and points of cut-off and compression provisionally settled, then it is possible to construct much more accurate diagrams with a view of determining how far the engine complies with the required conditions, and especially how far there is any waste expansion or fall of pressure between the cylinders.

In fig. 32, set off ab, the intermediate space R between the cylinders; ao, $b\ 5'$, the high- and low-pressure clearance volumes, c_a and c_b; and $o\cdot 5$, $\cdot 5'o'$ the volumes described by the high- and low-pressure pistons v_a and v_b. Draw the crank-pin semicircles on $o\cdot 5$, $\cdot 5'o'$.

On the vertical to the left take any distance to represent a revolution and divide it and the crank-pin circles into any number (conveniently 10) of equal parts. Number them in order, starting from the dead point. We can now draw the curves of sines or of piston displacement, numbered o, ·1, ·2, ..., points of which lie on the intersection of horizontal lines through the points of the revolution line and verticals through corresponding points of the crank-pin circles. This line has this property, that the horizontal abscissa from the vertical through a gives the whole volume of steam in cylinder and clearance space, at any given point of the revolution. The two lines of sines are so placed that for those parts of the revolution, during which both cylinders are in communication, the horizontal intercept between the two lines of sines is the total volume of steam in the two cylinders and receiver. Now select and mark on the lines of sines the cut-off and cushion-points for both cylinders; they are the points on the verticals through 1 and 5 of the high-pressure diagram, and through 11 and 14 of the low-pressure diagram. In the figure the high-pressure diagram is drawn for an initial pressure of 100 lbs. per sq. in., and the admission line o 1 is horizontal. 1 2 is an hyperbolic expansion-line for the assumed point of cut-off. The volumes are measured to the vertical through *a* in drawing this curve, because the steam is in the high-pressure cylinder and clearance. At 2 the high-pressure cylinder opens to the receiver and there is a drop of pressure (2 3), which will be determined presently. The pressure at 3 is the initial pressure at 1 o in the low-pressure cylinder. During 3–4 and 10–11 the steam expands in the high-pressure cylinder receiver and low-pressure cylinder, and its volume is the horizontal intercept between the two lines of sines. Thus the horizontal ·7–·7′ is the volume at those points of the revolution in the two cylinders. Calculating the pressure from the volume, it is to set it up on verticals through ·7 and ·7′ to give points on 3–4 and 10–11. At the point 4, corresponding to 11, the low-pressure slide-valve closes, and during 4–5 the steam is compressed in the high-pressure cylinder and receiver. The volumes for this part of the revolution are found by measuring from the curve of sines to the vertical through *b*. At 5 the cushion-point is reached and the steam is compressed in the high-pressure cylinder only, the volumes being now measured to the vertical through *a* in determining the pressures for the compression-line 5–6. In fact 5–6 is an hyperbola, of which the vertical through *a* is an asymptote.

To complete the low-pressure diagram, draw the back-pressure line 13–14 with an assumed value of the condenser pressure. 11–12 and 14–15 are hyperbolas, having the vertical through *b* as an asymptote ; that is, the volumes are measured from the low-pressure curve of sines to the vertical through *b*.

A curve of pressures in the receiver has also been drawn in the same way. The shading of the upper part of the diagram indicates what volumes of steam are in action

Proportions of Steam-Engine Cylinders 47

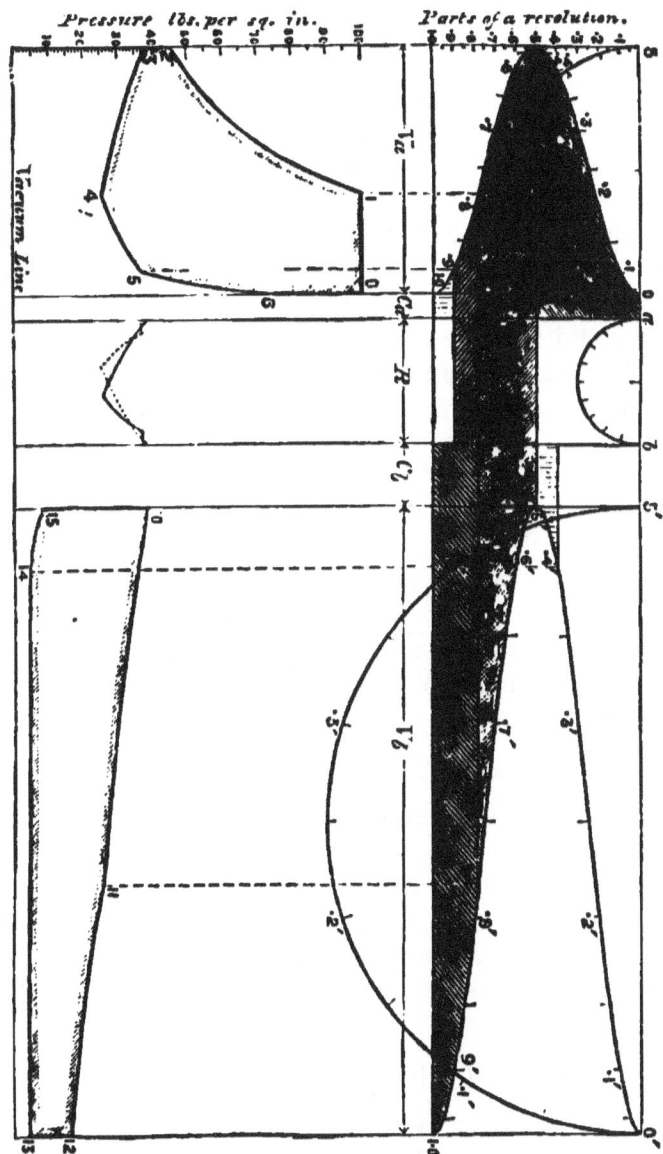

Fig. 32. - Cranks at 0° or 180°

at any given moment. The horizontal breadth of the shaded part is the steam volume.

Let p and v denote the pressure and volume of steam undergoing any operation of expansion and compression, then $pv =$ constant is the equation determining the indicator diagram curve. For any fraction of the stroke the values of p and v may be distinguished by subscript figures corresponding to those on the indicator diagrams in fig. 32. c_a and c_b are the high-pressure and low-pressure cylinder clearance volumes. R is the volume of the receiver.

At the end of admission the value of the product pv for the high-pressure cylinder is $p_1 (v_1 + c_a)$, and since the expansion is treated as hyperbolic,

$$p_2 (v_2 + c) = p_1 (v_1 + c_a) \quad . \quad . \quad . \quad (1)$$

which determines the pressure at the end of expansion. Release takes place at 2, and the steam in the cylinder mixes with that in the receiver. But the receiver was cut off in the previous stroke at 5, and hence the product pv for the steam in the receiver is p_5 R. Hence, during the period 3–4 the value of pv for the steam acting in the two cylinders conjointly is

$$p_1 (v_1 + c_a) + p_5 R = p_3 (v_3 + c_a + R + c_b). \quad . \quad (2)$$

an equation which determines p_3 when we have found the value of p_5.

The small quantity of steam p_1, c_b in the clearance space of the low-pressure cylinder is neglected for simplicity. The steam now expands in the diminishing high-pressure and increasing low-pressure cylinder volume, plus the clearances and receiver. Then

$$p_4 (v_4 + c_a + R + c_b + v_{11}) = p_1 (v_1 + c_a) + p_5 R \quad (3).$$

The low-pressure clearance and cylinder is now cut off, and during 4–5 the value of pv for the steam in the high-pressure cylinder and receiver is $p_4 (v_4 + c_a + R)$.

Hence

$$p_5(v_5 + c_a + R) = p_4(v_4 + c_a + R) \quad . \quad . \quad . \quad (4)$$

$$p_4 = p_5 \frac{v_5 + c_a + R}{v_4 + c_a + R};$$

putting this in 3, we get an equation which determines p_5. Then from (2) we can determine p_3. Thus all the points on the high-pressure diagram are fixed.

In the low-pressure cylinder during 10–11 the pressures are the same as on 3–4 for corresponding parts of a revolution, that is, for points on the same level on the two curves of sines. From 11–12 the value of $p\,v$ is $p_{11}(v_{11} + c_b)$, the volumes during expansion being measured from the low-pressure clearance line. At 12 there is release to the condenser. 13–14 is the condenser back-pressure line, and 14–15 an hyperbola for $p\,v = p_{14}(v_{14} + c_b) = p_{15}\,c_b$.

35. *Case II. Indicator diagram for a receiver engine. Cranks at 90°.*—Begin as in the previous case, setting off horizontally the high- and low-pressure cylinder volumes v_a and v_b, the clearance volumes c_a and c_b, and the receiver volume R. Take on the vertical through ·5 a length to represent a revolution and divide it into ten parts. Draw the crank-pin circles and divide them similarly, remembering that the points ·5 of the high-pressure crank and ·5′ of the low-pressure crank are at right angles; that is, at half a revolution, if the high-pressure crank is horizontal, the low-pressure crank is vertical. The numbering of corresponding points is then easy.

The high-pressure diagram (fig. 33) is drawn for an initial pressure of 140 lbs. per sq. in. The admission line 0–1 is horizontal; 1–2 is an hyperbolic expansion curve, the volumes being measured to the vertical through a. At 2 there is a drop as the receiver opens, which will be determined presently. During 3–4 the volume in the high-pressure cylinder is diminishing, and it is open to the receiver. The volumes are measured from the high-pressure curve of sines to the vertical through b. At 4, mid-stroke, there is a second fall of pressure as the low-pressure clearance opens, and this will be determined by calculation. At 5 the low-pressure cylinder opens to steam at mid-stroke of the high-pressure cylinder. The volumes for the curves 5–6 and 10–11 are the horizontal intercepts between the two curves of sines, the cylinders being in communication. 6 is the chosen cushion-point for the high-pressure cylinder, and 6–7 is an hyperbolic compression curve with the vertical through a as an asymptote.

For the low-pressure diagram, during 11–12, the cylinder is in communication with the receiver. The steam expands, the volumes being measured from the low-pressure curve of sines to the vertical through a. Now draw the back-pressure line 14–15. Then 12–13 and 15–16 are hyperbolic expansion and compression lines having the vertical through b as an asymptote, and the volumes are measured from the curve of sines to this line.

As before, a curve of receiver pressure has been drawn, and the shading of the upper part of the figure indicates what volumes of steam are acting in either cylinder at any moment.

To determine the points on the diagram we have the following equations. During 1–2, expansion in high-pressure cylinder,

$$p_1(v_1 + c_a) = p_2(v_2 + c_a) \quad \ldots \quad (5)$$

During 3–4, compression in high-pressure cylinder and receiver,

$$p_3(v_3 + c_a + R) = p_4(v_4 + c + R) \quad \ldots \quad (6)$$

At 4, midstroke of the H. P. cylinder, the L. P. cylinder opens, and there is a drop from the steam filling the clearance space of L. P. cylinder, in which the product pv has the value $p_{16} c_b$. Then

$$p_{16} c_b + p_4(v_4 + c_a + R) = p_5(v_4 + c_a + R + c_b) \quad (7)$$

During 5–6 and 10–11

Proportions of Steam-Engine Cylinders 51

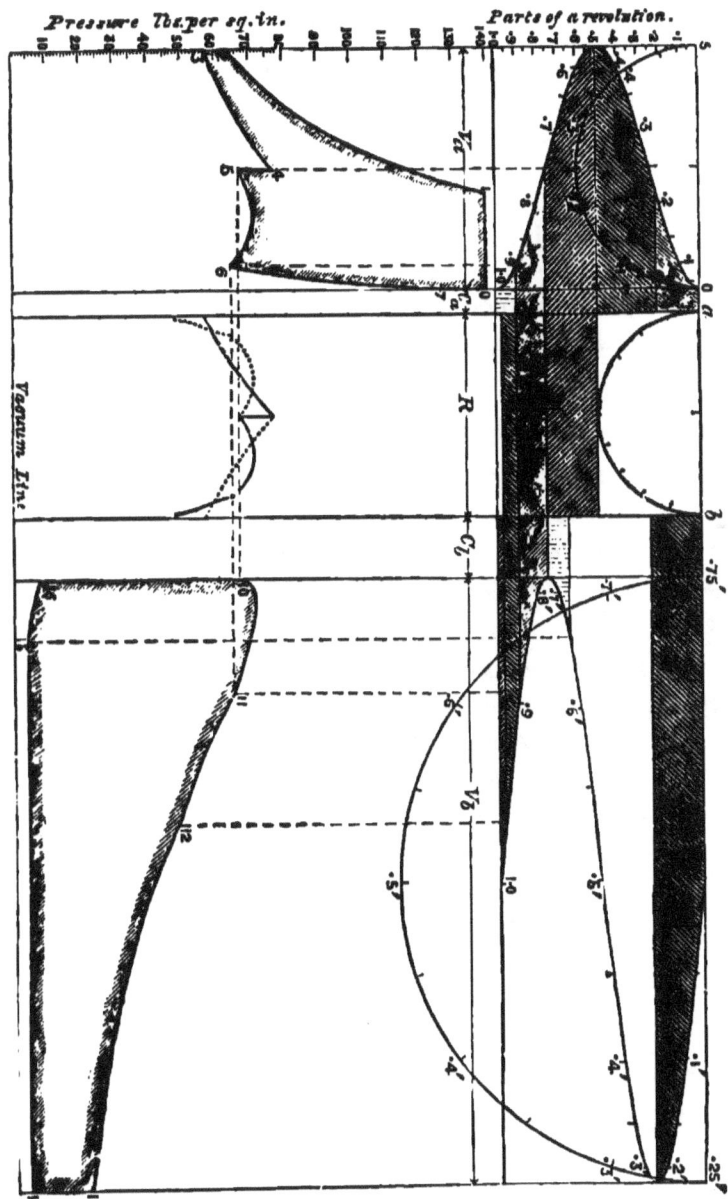

Fig. 33—Cranks at right angles

$$p_6(v_4 + c_a + R + c_b) = p_6(v_6 + c_a + R + c_b + v_{11}) \quad (8)$$

At 6 compression begins in the H. P. cylinder, and 6-7 is an hyperbola with the vertical through a as asymptote. The H. P. diagram can now be completed; for if the L. P. back-pressure line 14-15 is drawn, and 15-16 as an hyperbola with the vertical through b as an asymptote, the quantity $p_{16} c_b$ is determined.

To complete the L. P. diagram. During 11-12 the steam expands in receiver and L. P. cylinder,

$$p_6(R + c_b + v_{11}) = p_{12}(v_{12} + c_b + R) \quad . \quad (9)$$

At 12 cut off in L. P. cylinder occurs, thence the steam expands in L P. cylinder only.

$$p_{12}(v_{12} + c_b) = p_{13}(v_{13} + c_b) \quad . \quad . \quad . \quad (10)$$

For the drop of pressure in the H. P. cylinder we have the equation, since p_{12} is the pressure in receiver when expansion into the L. P. cylinder ceases,

$$p_2(v_2 + c_a) + p_{12}R = p_3(v_2 + c_a + R) \quad . \quad . \quad (11)$$

The simultaneous equations are a little troublesome, but only require patience for solution.

For a triple engine a similar process may be used, two pairs of diagrams being drawn, one for the H. P. and I. P. cylinders, and one for the I. P. and L. P. cylinders.

CHAPTER III

LINKWORK—CRANKS AND ECCENTRICS

36. According to the most modern views of the nature of a machine, it consists of a series of pairs of elements linked together into one or more kinematic chains. The cylindrical journal and its pedestal form at once the commonest and most typical example of a pair of elements. If the pedestal is fixed, the journal can rotate and can have no other motion. A pair of elements in which the relative motion is thus fixed is called a 'closed pair,' and in machinery the only pairs which are useful are pairs which are closed either geometrically or virtually by constraints which prevent any relative motion other than the definite relative motion required. Suppose a series of pairs taken and linked, one element of one pair to one element of another pair. We get a kinematic chain. Such a chain is termed a 'closed chain,' if any one link being fixed all the other links have one definite motion, and one only, relatively to the fixed or frame-link. Obviously, only closed chains are useful in machinery, because definite motions are to be given to the working tools or other driven parts of the machine.

The simplest of all kinematic closed chains consists of four cylinder pairs (or journals and bearings) linked together, as shown in fig. 34, the fixed link being shaded. As drawn, b rotates and c oscillates relatively to the frame-link; d has a more complex motion. Links like b or c are termed 'cranks' or 'levers,' and links like d, the bearings of which do not rest on the frame but on other moving pieces, and which have in general a more complex motion than pieces resting on the frame, are called 'connecting' or 'coupling

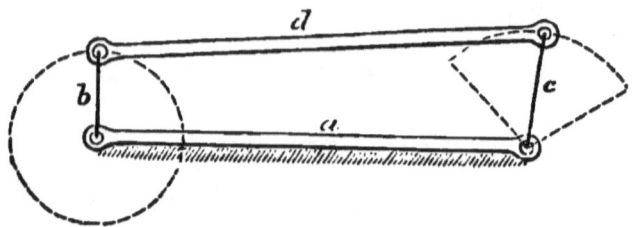

Fig. 34

rods.' The chain shown in fig. 34 may be recognised as being identical with the half-beam c, the crank b, connecting rod d and column and entablature a of a beam engine.

Such a simple closed chain is capable of various transformations. The relative size of the parts may be changed sometimes with important results. If the fixed link is changed, the chain is said to be *inverted*, and then, although the relative motion of each pair of links is unchanged, the general motion of the whole mechanism is often quite different. Thus, if the short link b, fig. 34, is fixed, we get two completely rotating cranks linked together (see b, fig. 35). The mechanism is identical with that of the drag-link coupling used to connect two shafts not exactly in line. If the link c, fig. 34, is fixed, we get two levers, having only a limited range of oscillation (see c, fig. 35). Some forms of Watt's parallel motion are identical with this chain (d, fig. 35).

37. Suppose that now the chain of fig. 34 is changed by substituting a slide block and slide bars for one pair of cylindrical elements. Then we get a chain (fig. 36) exceedingly common in machinery, and easily recognisable as identical with the direct-acting engine mechanism. Here a rotating crank, b, gives straight-line motion to the slide-block c and crosshead, or *vice versâ*. There are three cylinder pairs and one sliding pair. If this chain is inverted,

we get, by fixing the link d, a mechanism identical with that of the oscillating steam engine (fig. 35, g); by fixing b,

INVERSION OF CHAINS

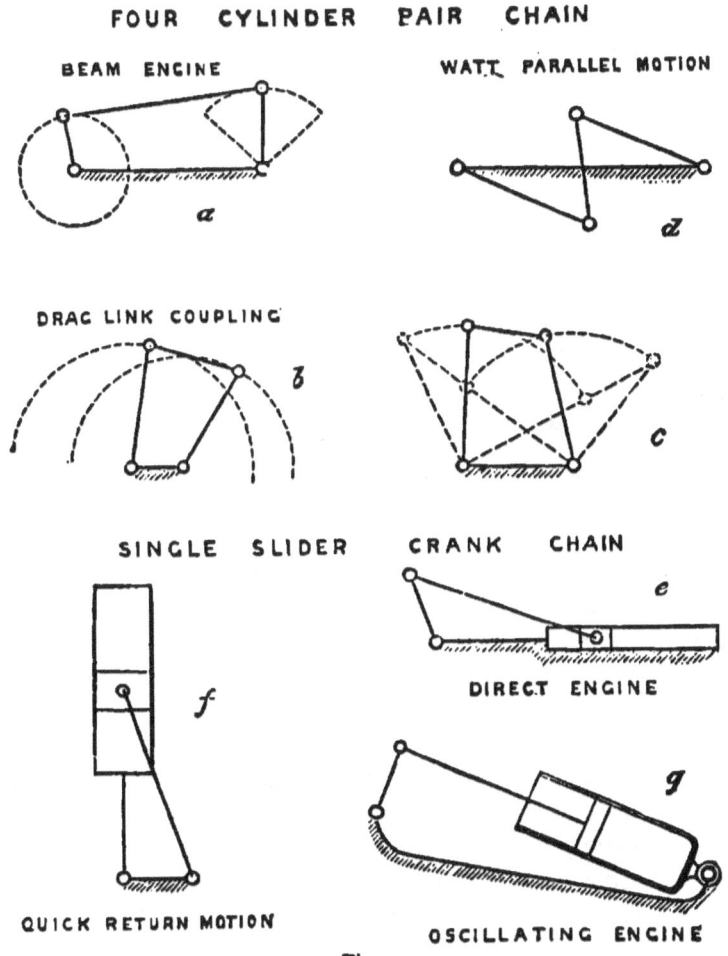

Fig. 35

a mechanism used at one time to get a slow forward and quick return motion for planing machines (fig. 35, f); and by fixing c, a mechanism which has been utilised in the pendulum steam pump. This chain is known as the 'single-slider crank chain.'

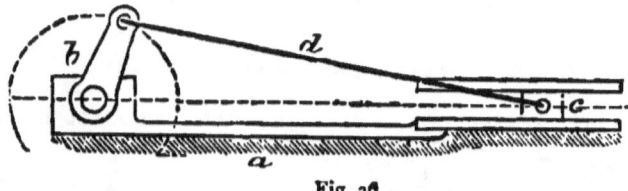

Fig. 36

By replacing two of the cylinder pairs in fig. 34 by sliding pairs we get the double-slider crank chain shown in

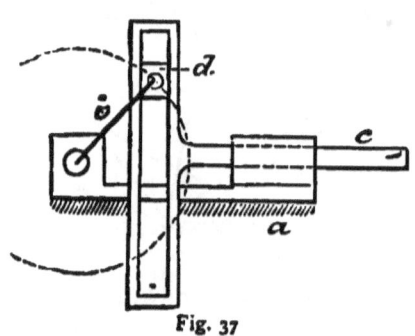

Fig. 37

fig. 37, recognisable as identical with some forms of steam-pump, and this again takes other forms by inversion.[1]

38. *Piston displacement.*—Let $e\,d\,g$ fig. 38, be the crank-pin circle, and dc the connecting-rod length. If we take $ea = gb = dc$, then ab is the path of the crosshead, and this for all practical purposes is the path of the piston, which is rigidly attached to the crosshead. If od is any position of the crank, then c is the corresponding position of the crosshead, found by striking an arc with d as centre and dc as radius. Since $eg = ab$, it is often convenient to take eg as representing the path of the piston. Then f is the piston or crosshead position corresponding to c, and is found by striking an arc with centre c and radius cd. Obviously f divides eg in the same ratio that c divides ab, and $fg = cb = x$ is the displacement of the piston from

[1] A much fuller account of Reuleaux's analysis of machines will be found in Cotterill's 'Applied Mechanics' and Kennedy's 'Mechanics of Machinery.'

the beginning of the stroke. Draw dh perpendicular to eg. Then fh is in general not large compared with the stroke.

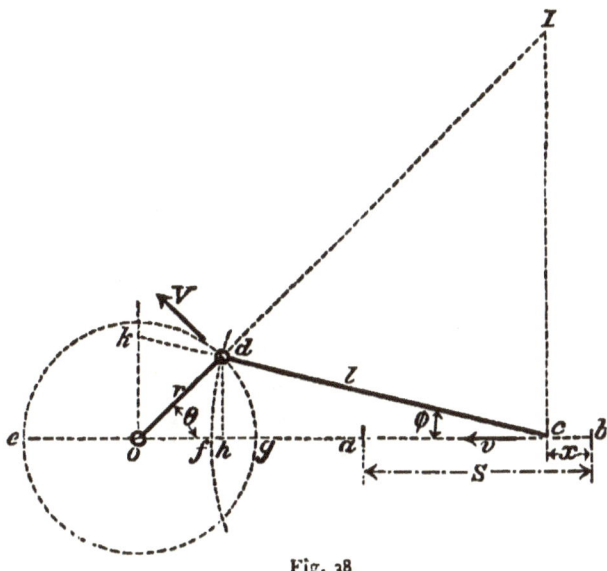

Fig. 38

If dc were indefinitely large, f would coincide with h, and the distance fh would vanish. Hence fh may be called the error of displacement due to the obliquity of the connecting rod. In double-slider crank chains, which are kinematically identical with single-slider crank chains having indefinitely long connecting rods, h moves exactly in the same way along eg that the reciprocating parts move along their stroke path. In many problems about single-slider crank chains it is a sufficient approximation to neglect the distance fh, and to take h as corresponding to the position of the piston or crosshead when eg is the stroke.

If r is the crank radius o d, l the connecting-rod length cd, $\rho = \dfrac{r}{l}$, the ratio of crank to connecting rod, which is usually between $2/7$ and $1/5$; θ the angle the crank makes with the line of stroke

$$x = r(1-\cos\theta) \pm l\left[1 - \sqrt{1-\rho^2 \sin^2\theta}\right]$$

where the $+$ sign is taken if θ is measured from the inner and the $-$ sign if θ is measured from the outer dead point, x being measured from the dead point. For $\theta = 90°$, $x = r \pm \frac{1}{2}\frac{r^2}{l}$. If $\rho = 0$, $x = r(1-\cos\theta)$.

39. Piston or crosshead velocity.—In a number of cases the velocity, V, of the crank-pin in the direction of its path may be treated as constant, and then the velocity, v, of the crosshead varies. Draw $c\,1$, $d\,1$, perpendicular to the directions of motion of c and d, then 1 is the instantaneous axis of rotation of the connecting rod, the point about which the connecting rod is rotating at the moment. Hence $v/\text{V} = c\,1/d\,1$. Produce the connecting rod to meet the perpendicular to the line of stroke in k.[1] Then from the similarity of triangles $o\,k\,d$, $d\,1\,c$, $v/\text{V} = o\,k/o\,d$. If to any scale $o\,d$ represents the crank-pin velocity, $o\,k$ represents to the same scale the piston velocity. Obviously

$$v = \text{V}\,\frac{\sin(\theta + \phi)}{\cos \phi}$$

or approximately

$$v = \text{V}\sin\theta\,(1 \pm \rho \cos\theta),$$

and if the obliquity of the connecting rod is neglected,

$$v = \text{V}\sin\theta.$$

Graphic representation of piston or crosshead velocity.—Let $o\,d$, fig. 39, be the crank length r, and $d\,c$ the connecting-rod length l, as before. Take $df = \text{V}$, the crank-pin velocity, draw fg parallel to the connecting rod to meet a, perpendicular to the line of stroke in g. Then cg is the crosshead or piston velocity v. If the ordinate, cg, is found for several positions of the crank, the locus of g is an oval curve, which becomes an ellipse if the connecting rod

---

[1] The construction is still correct if the line of stroke does not pass through o.

is indefinitely long. This is the curve of piston velocity with *abscissa* = piston displacement, and ordinate = piston

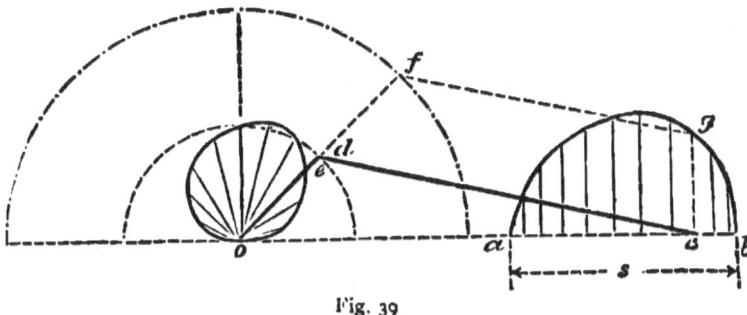

Fig. 39

velocity. It is sometimes convenient to set off $cg = v$ on the crank at oc. Then the locus of e is a kind of lemniscate curve which becomes a pair of circles for the case of a very long connecting rod. This is the polar curve of piston velocities, the vector of the curve in the direction of the crank being the piston velocity. The maximum piston velocity occurs very nearly when the crank and connecting rod are at right angles.

40. *Crank and connecting rod. Forces acting at crosshead and crank pin.*—In steam engines the crank is driven by the steam pressure on the piston transmitted to it through a connecting rod. If *effort* is defined to be the component of the forces acting at a joint or pair of elements, which is in the direction of motion, and if, further, the forces due to inertia of the moving pieces and the small frictional resistances are neglected, then the principle of conservation of energy shows that the efforts at two connected joints are inversely as their velocities.

If p is the effective steam pressure on the piston in lbs. per sq. in. (that is, the excess of pressure on one side over back pressure on the other) and A the area of the piston in sq. in., then $P = p A$ is the total effort at the piston. We may call p the effort reckoned per unit of area of piston, and in what follows the other efforts will for convenience be reckoned in the same way; that is, they must be multiplied by the piston area to get the total effort. If R is the crank radius, the crank pin travels 2π R in one revolution while the piston travels 4 R. Hence, the *mean effort* on the crank pin is less than the mean effort on the piston in the ratio of 4 : 2π or $2 : \pi$. If p_m is the mean effective steam pressure during a revolution, the mean crank-pin effort is $t_m = 2 p_m/\pi$, both reckoned per unit of piston area.

Since the crank radius is constant, the twisting moment on the crank shaft is proportional to the effort at the crank pin, and the mean twisting moment is $t_m R = 2 p_m R/\pi$.

In determining the strength of the machine parts, it is not enough to know the mean efforts and moments; it is necessary to examine the forces transmitted in all positions of the mechanism.

Dealing first with the connecting rod, let L be the length of connecting rod, and put $n = L/R$. Then n in actual engines is rarely less than $3\frac{1}{2}$ or more than 6, though in some special machines it has a smaller value. If β is the angle B A C in fig. 40, then the piston pressure P acting on the crosshead A is balanced by a pressure P tan β normal to the slidebars and a thrust P sec β along the connecting rod. Reckoned per unit of piston area, these forces are p tan β and p sec β. Resolving the connecting-rod thrust at the crank pin B into an effort t tangential to the crank-pin circle and a radial thrust r along the crank, we get

$$t = p \frac{\sin(\theta + \beta)}{\cos \beta}$$
$$r = p \frac{\cos(\theta + \beta)}{\cos \beta}$$

where t and r are again reckoned per unit of piston area. These expressions are not very convenient.

41. *Approximate treatment of the forces acting on crank pin. Steam pressure assumed constant and obliquity of con-*

necting rod neglected.—Let s s (fig. 40) represent on any scale the stroke of the engine, A C, to any scale the steam pres-

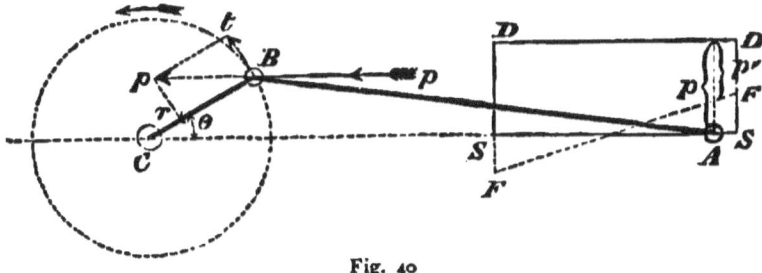

Fig. 40

sure *p* per sq. in., which for simplicity is assumed constant through the stroke. Then the indicator diagram is the rectangle S D D S. Let C B, B A, be the position of crank and connecting rod at any instant. For more simplicity, suppose the obliquity of the connecting rod neglected so that the thrust of the connecting rod, *p* per unit of piston area, is taken to be acting at B in the direction of the arrow parallel to the line of stroke A C. The pressure *p* at B is balanced by the tangential resistance *t* of the crank to rotation and by the radial thrust *r* along the crank. If, therefore, B *p* is taken equal to *p* and the parallelogram of forces completed, it is seen that

$$t = p \sin \theta$$
$$r = p \cos \theta$$

where θ is the angle through which the crank has turned from the beginning of the stroke. The values of *t* and *r* vary as the crank turns, but the expression for them under the stated restrictions is very simple.

Now suppose the construction shown in fig. 40 repeated for several positions of the crank. In fig. 41 take *o b* parallel to the crank C B in fig. 40, and along this measure *o t*=the tangential effort B *t*. All the points *t* corresponding to different crank positions will lie on the circumference of the full circle $o\,t\,t'$, and for the return stroke on the lower full circle. These two circles, therefore, form a polar diagram of crank-pin effort, the vector parallel to any crank position being the tangential effort for that position. Similarly, if we take $o\,r =$ the radial thrust B r and repeat the process for other crank positions, the points r for the forward and return strokes will lie on the dotted circles, which are polar diagrams of radial thrust.

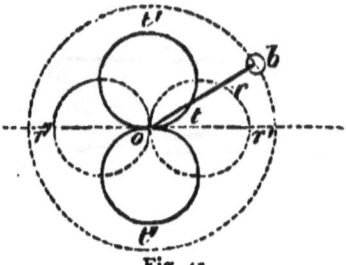

Fig. 41

The twisting moment on the crank shaft is equal to t R or B $t \times$ C B in fig. 40. But as the crank radius is constant, the twisting moment is proportional to B $t = t$. Hence the full circles in fig. 41 also represent polar diagrams of crank shaft twisting moment.

The polar curves are easily drawn directly without going through the construction shown in fig. 40, and give in a very convenient way the variation of crank-pin effort, radial thrust and twisting moment for every position of the mechanism. The simplicity arrived at, however, is due to assuming the steam pressure constant and neglecting inertia and the connecting-rod obliquity.

42. *Correction of approximate diagram of piston and crank-pin effort for inertia.*—There is, however, one factor which greatly modifies the forces at the crank pin. In engines, and especially in high-speed engines, the inertia of the heavy parts connected with the crosshead, and reciprocating with it, greatly alters the distribution of the crank effort during the stroke, although the mean value for a semi-revolution is not altered. During the earlier half of the stroke the velocity of all horizontally moving pieces is accelerated and during the later half retarded. Part of the steam pressure is used in accelerating the heavy reciprocating parts

in the earlier half of the stroke, and during the later half the pressure required to retard them forms virtually an addition to the steam pressure.

Suppose, as before, the obliquity of the connecting rod neglected, and that we have only to do with a horizontal variation of the velocity of the moving pieces.

In steam engines the variation of crank-pin velocity is very small. Let v be the constant velocity of the crank pin and R the radius of the crank. Then for the position C B of the crank (fig. 40), the horizontal velocity of the crosshead A is $v \sin \theta$. Then the acceleration horizontally of the connecting rod, crosshead and piston is

$$\frac{d(v \sin \theta)}{dt} = \frac{d(v \sin \theta)}{d\theta} \cdot \frac{d\theta}{dt}$$

$= v \cos \theta \times \omega$, where ω is the angular velocity of the crank; or putting $\omega = \frac{v}{R}$, the acceleration is $\frac{v^2 \cos \theta}{R}$. Now let W be the weight of the horizontally moving parts in lbs., $\frac{W}{g} \cdot \frac{v^2 \cos \theta}{R}$ is the whole force required to accelerate them. Or if w is the weight of the horizontally moving parts in lbs. per sq. in. of piston, the accelerating force, in lbs. per unit of piston area, is —

$$\frac{w}{g} \cdot \frac{v^2 \cos \theta}{R},$$

which for $\theta = 0°$, or $180°$ becomes

$$\pm \frac{w}{g} \cdot \frac{v^2}{R}.$$

For any position A of the crosshead, when the crosshead has moved a distance,

$$x = A\,S = R(1 - \cos \theta)$$

the ordinate of a curve representing the accelerating force is

$$v = \frac{w}{g} \cdot \frac{v^2 \cos \theta}{R}$$
$$= \frac{w}{g} \cdot \frac{v^2 (R-x)}{R^2}$$

the equation to a straight line. Hence, if S F, S F' are taken equal to $\pm \frac{w}{g} \cdot \frac{v^2}{R}$ on the same scale as that used for the steam pressure, and F F' is joined, then the effective force at the crosshead driving the crank, reckoned per unit of area of piston, is the vertical ordinate p' of the figure F D D F'. In this case the curves of tangential and radial pressure are no longer circles.

It is worth while to get at the forces acting at the crank pin in the simple way here adopted as a first step. But obviously very arbitrary assumptions have been introduced. For practical purposes the forces acting must be determined more accurately, whether we are considering the dynamical action of the engine or require to calculate the strength of the parts of the mechanism. If the assumptions made above are abandoned, the algebraical expressions for the forces acting at the crank pin become very complex. But they can be determined graphically without difficulty.

43. *Forces acting at the crank pin when the steam pressure varies during the stroke and the obliquity of the connecting rod is taken into account.*—The first restriction to get rid of is the neglect of the varying obliquity of the connecting rod. Still leaving on one side the influence of the inertia of the parts and the friction, the efforts at two joints are inversely as their velocities. Hence the methods already described for finding the velocity ratio of two joints serve also for determining the ratio of the efforts. Let C B, B A (fig. 42), represent as before any position of the crank and connecting rod. Let S D D S be an indicator diagram, S S being the stroke of the crosshead, and S D the initial steam pressure. Produce C B, and a perpendicular at A to S S, to meet in I.

Then I is the point about which the link B A is rotating at the instant considered. If V is the velocity of B and v the velocity of A,

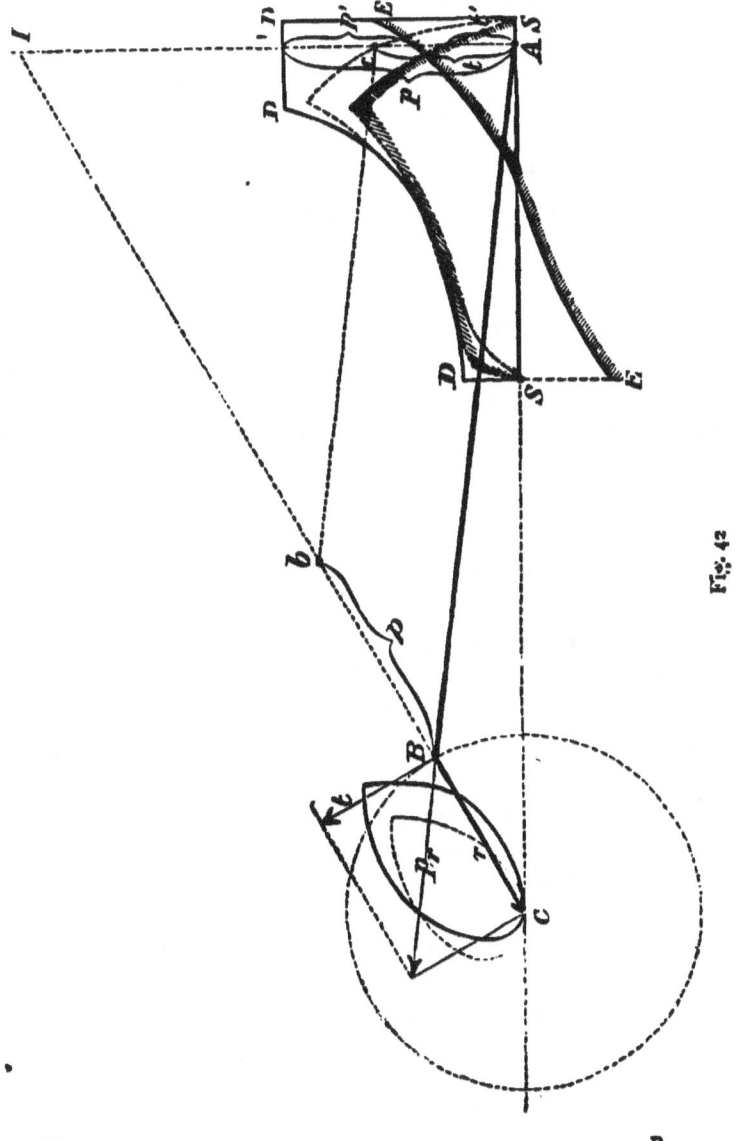

Fig. 42

$$\frac{v}{V} = \frac{AI}{BI}.$$

Let p, the ordinate of the indicator diagram, be the pressure acting at A, and t the tangential pressure at B; then

$$pv = tV$$

$$\frac{t}{p} = \frac{AI}{BI}.$$

Take $Bb = p$; draw bc parallel to BA,

$$\frac{t}{p} = \frac{AI}{BI} = \frac{cA}{bB};$$

$$\therefore \quad t = cA.$$

Values of t found for all points of the stroke give the dotted curve scs. If the values of t are laid off on the corresponding positions of the crank, we get the oval curve shown by a full line on the left, which corresponds to one of the full circles in fig. 41. The return stroke would give a similar curve below the horizontal line.

Set off t tangential to the crank-pin circle at B, and complete the parallelogram of forces. We thus get the radial component r of the crank-pin pressure, and the resultant pressure p_r along AB on the crank pin.

If this construction is made for several positions of the crank, it will be seen that the tangential component t vanishes when the crank is at the dead point, and, except when the pressure on the piston diminishes before half-stroke, it reaches its maximum when the crank and connecting rod are at right angles. Its maximum value, if R = crank radius and L = connecting-rod length, is,

$$t_{max} = p \frac{\sqrt{R^2 + L^2}}{L} \quad . \quad . \quad . \quad (1)$$

Since L is usually 4 to 5 times R,

$$t_{max} = 1\cdot02 \text{ to } 1\cdot03\, p \quad . \quad . \quad . \quad (2).$$

The radial component r vanishes when the crank and

connecting rod are at right angles and is greatest and equal to p when the crank is at the dead point.

44. *Curve of force due to inertia of reciprocating parts, when the obliquity of connecting rod is taken into account.*—In fig. 43, s s represents the stroke, and $a\,a$, determined as in §42, is the curve of force due to inertia for the case of an

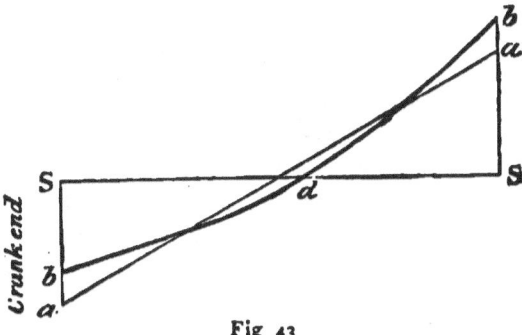

Fig 43

indefinitely long connecting rod. When the connecting rod is short (say $3\frac{1}{2}$ to 5 crank lengths) the inertia curve takes a form like $b\,b$, if we treat the reciprocating parts as having a horizontal motion identical with that of the crosshead. The piston, piston-rod and crosshead all move horizontally together, but the connecting rod has a horizontal motion a little different. However, the error of the assumption above is not important unless the connecting rod is very short.

To determine the curve $b\,b$, various methods are available. For instance, if the curve of crosshead velocity is first drawn, the principle explained in I. § 22, p. 37, may be used. The subnormal of the velocity curve at any point is the acceleration at that point. For most practical purposes, however, it is sufficient to determine the three points $b\,d\,b$ and draw a flat curve through them, and this is very easily done.

At the ends of the stroke the force due to inertia has already been found to be $\pm \dfrac{w}{g}\dfrac{v^2}{R}$, reckoned per unit of area of piston and neglecting the obliquity of the connecting rod. When the connecting rod is n cranks in length, this expression becomes, as will be shown presently,

$$+ \frac{w}{g} \frac{v^2}{R} \left(1 + \frac{1}{n}\right)$$

for the inner dead point, and

$$- \frac{w}{g} \frac{v^2}{R} \left(1 - \frac{1}{n}\right)$$

for the outer dead point.[1] Set off sb, sb, equal to these values. The point d is the position of the crosshead when the connecting rod and crank are at right angles, which is easily found by construction. The three points $b\,d\,b$ through which the curve is to be drawn are therefore determined. So long as $n > 4$ these three points determine the inertia curve accurately enough for most practical purposes.

In fig. 42 a curve of force due to inertia E E has been drawn to the same scale as the indicator diagram. Then the effective horizontal force at the crosshead driving the crank, estimated per unit of piston area, is the intercept p' between D D D and E E. Using this instead of p, and using the method shown in fig. 42, we get t' as the value of the crank-pin effort after allowing for inertia. Values of t' found for successive positions of the mechanism give the shaded curve of crank-pin efforts, and these, set off on the crank positions, give the dotted polar curve (on the left) of crankpin efforts or twisting moments.

45. *Exact determination of the influence of inertia on the diagram of crank-pin effort.*—The most important effect of inertia arises from the resistance to acceleration of the reciprocating masses when considered as having a translational velocity v parallel to the line of stroke. But, besides this, the connecting rod exerts a centrifugal pull on the

[1] See also Kennedy, 'Mechanics of Machinery,' p. 355.

crosshead pin and a resistance to angular acceleration. An exact determination of the forces thus arising was given by Prof. Fleeming Jenkin ('Trans. Roy. Soc.' Edinburgh, vol. xxviii. p. 711).

A method most convenient for arithmetical calculation will be given first, and then a graphic method generally more convenient for practical purposes.

Let $AB = R$ be the crank radius; $BC = L$ the connecting-rod length, G the centre of gravity of the connecting rod, and H its centre of percussion. Then if $CG = L_0$ and k is the radius of gyration of the connecting rod, $CH = k^2/L_0$.

Let V = velocity of crank pin B and v the velocity of crosshead C, θ = angle BAC, α = angle BCA, and let t be

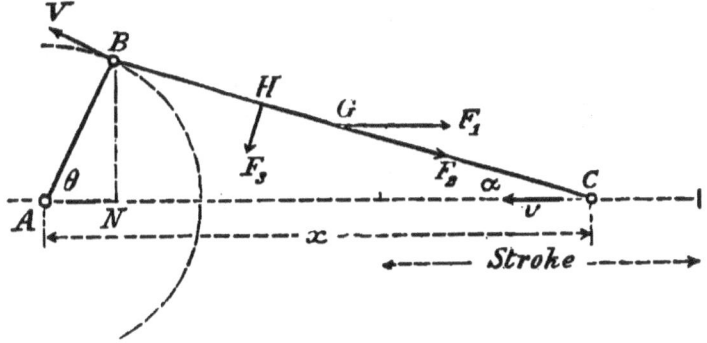

Fig. 44

the time from the beginning of the stroke. The angular velocity of the crank, which is taken to be constant, is $\omega = V/R = d\theta/dt$. The angular velocity of the connecting rod is $d\alpha/dt$ and its angular acceleration $d^2\alpha/dt^2$. The translational velocity of the reciprocating parts is $v = dx/dt$, and the translational acceleration is $dv/dt = d^2x/dt^2$. For simplicity put $L/R = n$ or $L = nR$.

Now $\alpha = \sin^{-1}\dfrac{BN}{BC} = \sin^{-1}\left(\dfrac{\sin\theta}{n}\right)$

$$\dfrac{d\alpha}{dt} = \dfrac{\cos\theta}{\sqrt{(n^2-\sin^2\theta)}} \cdot \dfrac{d\theta}{dt} \qquad . \qquad . \qquad (3)$$

Differentiating again and remembering that $d^2\theta/dt^2 = 0$ we get

$$\frac{d^2 a}{dt^2} = -\frac{\sin\theta(n^2-1)}{(n^2-\sin^2\theta)^{3/2}}\left(\frac{d\theta}{dt}\right)^2 \qquad . \qquad . \quad (4)$$

But $x = AC = AN + NC$
$= R\cos\theta + L\cos a$

$$v = \frac{dx}{dt} = -R\sin\theta\frac{d\theta}{dt} - L\sin a\frac{da}{dt}$$

$$\frac{dv}{dt} = \frac{d^2 x}{dt^2} = -R\cos\theta\left(\frac{d\theta}{dt}\right)^2 - L\cos a\left(\frac{da}{dt}\right)^2$$
$$- L\sin a\frac{d^2 a}{dt^2}.$$

Substituting the values above and putting $R\sin\theta$ for $L\sin a$ and $R\sqrt{(n^2-\sin^2\theta)}$ for $L\cos a$, we obtain finally

$$f = \frac{dv}{dt} = \frac{d^2 x}{dt^2} = -R\left(\frac{d\theta}{dt}\right)^2$$
$$\left\{\cos\theta + \frac{n^2\cos 2\theta + \sin^4\theta}{(n^2-\sin^2\theta)^{3/2}}\right\}$$
$$= -R\left\{\cos\theta + \frac{n^2\cos 2\theta + \sin^4\theta}{(n^2-\sin^2\theta)^{3/2}}\right\}\omega^2 \qquad . \qquad . \quad (5)$$

Let $f = K R \omega^2$. Then K has the following values. The signs are given so that $+$ values correspond to forces to be added to the steam pressure on the piston and $-$ quantities to forces to be deducted from the steam pressure, in finding the effective thrust or pull at the crank pin.

Forward stroke from inner dead point,

Angle	K
0°	$-\left(1+\dfrac{1}{n}\right)$
45°	$\cdot\left\{\cdot 707 + \dfrac{0\cdot 25}{(n^2-0\cdot 5)^{3/2}}\right\}$
90°	$\dfrac{1}{\sqrt{n^2-1}}$

Linkwork—Cranks and Eccentrics

Angle	K
135°	$\cdot 707 - \dfrac{0 \cdot 25}{(n^2 - 0 \cdot 5)^{3/2}}$
180°	$1 - \dfrac{1}{n}$

Return stroke from outer dead point.

Angle	K
180°	$-\left(1 - \dfrac{1}{n}\right)$
225°	$-\left\{ \cdot 707 - \dfrac{0 \cdot 25}{(n^2 - 0 \cdot 5)^{3/2}} \right\}$
270°	$-\dfrac{1}{\sqrt{n^2 - 1}}$
315°	$\cdot 707 + \dfrac{0 \cdot 25}{(n^2 - 0 \cdot 5)^{3/2}}$
360°	$1 + \dfrac{1}{n}$

In the return stroke the values are those of the forward stroke in reverse order. From these equations the values of f for five points of the stroke are easily calculated for any ratio n of connecting rod to crank.

Let w_1 be the mass in lbs. of the piston, crosshead and other parts which have a simple motion of translation, and w_2 the mass in lbs. of the connecting rod, which will be regarded for the moment as having a simple motion of translation also. Let w_1 w_2 be the same quantities reckoned per unit of piston area. Then the total resistance to translational acceleration is

$$(w_1 + w_2)f/g \quad \text{or} \quad (w_1 + w_2)f/g,$$

according as we want the total force or the force per unit of piston area. This acts away from the centre of the crank shaft if $+$ and towards it if $-$, and is to be deducted from or added to the piston effort in estimating the thrust transmitted to the crank pin.

But the connecting rod has not a simple motion of translation. It may be regarded as having a motion of translation with velocity v parallel to the line of stroke, combined with a rotation about the crosshead pin. Hence the force due to acceleration is the resultant of three components: (1) a force F_1 (fig. 44) required for the translational acceleration acting at G parallel to the line of stroke, and which has been included in the expression above. (2) A force F_2 equal and opposite to the centrifugal force and acting along the rod towards the crosshead. (3) A force F_3 required for angular acceleration acting at H at right angles to the rod.

The values of these forces can be at once written down.

$$F_1 = \frac{W_2}{g} \frac{d^2 x}{d t^2}$$

$$F_2 = \frac{W_2}{g} L_0 \left(\frac{d a}{d t}\right)^2$$

$$F_3 = \frac{W_2}{g} L_0 \frac{d^2 a}{d t^2}$$

If these are to be taken into account they must be separately calculated and combined graphically.

46. *Exact graphic construction of curve of forces due to translational inertia of reciprocating parts.*—The simplest direct construction for determining the forces due to translational inertia is that of Rittershaus ('Civilingenieur,' xxv. s. 461). The acceleration of C is found as the difference of F E and E A, which are obtained by producing the connecting rod to meet a perpendicular through A to the line of stroke in D; then drawing D E perpendicular to D C and D F parallel to B E.

In fig. 45 this construction is given for the case of an exceptionally short connecting rod, to show in a marked way the alteration of form of the curve as the ratio L/R diminishes.

Let A B be the crank, B C the connecting rod of an ordinary horizontal engine, fig. 45. Produce the crank so that A Z is perpendicular to B's direction of motion, and draw C Z perpendicular to the line of stroke and to C's direction of motion. Then Z is the instantaneous axis of the link BC, and, as is well known, putting V for B's velocity and v for C's velocity,

$$\frac{v}{V} = \frac{CZ}{BZ}.$$

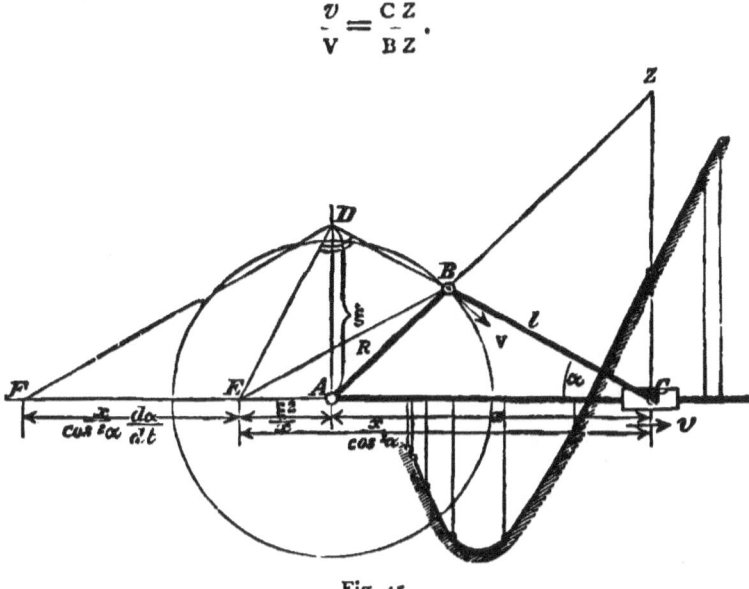

Fig. 45

Produce CB to meet a perpendicular AD to the line of stroke AC. Call R the crank radius AB, ξ the intercept AD. Then from similar triangles

$$\frac{v}{V} = \frac{\xi}{R}$$

$$v = \left(\frac{V}{R}\right)\xi.$$

Draw DE perpendicular to DC, and DF parallel to BE. Let AC = x and the angle ACB = a. Then

$$\xi = x \tan a,$$

differentiating with respect to the time

$$\frac{d\xi}{dt} = \left(\frac{R}{V}\right)\frac{dv}{dt} = \frac{x}{\cos^2 a}\frac{da}{dt} + \tan a \frac{dx}{dt}.$$

But
$$\frac{dx}{dt} = v = \left(\frac{V}{R}\right)\xi$$
and the acceleration f is
$$f = \frac{dv}{dt} = \frac{V}{R}\frac{x}{\cos^2 a}\frac{da}{dt} + \left(\frac{V}{R}\right)^2 \xi \tan a.$$
The angular velocity of the connecting rod round $c = \frac{da}{dt} = -\frac{V}{BZ}$. But $BZ = R \cdot \frac{BC}{BD}$. Hence
$$\frac{da}{dt} = -\frac{V}{R}\frac{BD}{BC};$$
consequently,
$$f = \frac{dv}{dt} = -\left(\frac{V}{R}\right)^2 \left\{ \frac{x}{\cos^2 a}\frac{DB}{BC} - \xi \tan a \right\},$$
or for the scale which makes $V = R$
$$\frac{dv}{dt} = -\left\{ \frac{x}{\cos^2 a}\frac{BD}{BE} - \xi\frac{\xi}{x} \right\}.$$
But by construction
$$EC = \frac{DC}{\cos a} = \frac{x}{\cos^2 a}$$
$$\frac{FE}{EC} = \frac{DB}{BC};$$
$$FE = \frac{x}{\cos^2 a}\frac{BD}{BC};$$
also
$$\frac{EA}{\xi} = \frac{\xi}{x}$$
$$EA = \frac{\xi^2}{x};$$
consequently,
$$f = -\{FE - EA\}.$$

For the dead points, at which the construction fails, since $a = 0$, we have for the inner dead point
$$x = l + R \text{ and } \frac{DB}{BC} = \frac{R}{l};$$

for the further dead point

$$x = l - \text{R} \text{ and } \frac{DB}{BC} = -\frac{R}{l}.$$

Hence, for the inner dead point, on the scale for which $v = R$,

$$f_1 = \frac{dv}{dt} = -(l+\text{R})\frac{R}{l} = -\text{R}\left(1 + \frac{R}{l}\right)$$

and for the outer

$$f_2 = \frac{dv}{dt} = (l-\text{R})\frac{R}{l} = \text{R}\left(1 - \frac{R}{l}\right).$$

Let w be the weight of the horizontally moving parts per unit of piston area. The whole accelerating force, per unit of piston area, is

$$\frac{wf}{g};$$

we have therefore only to measure the values of f on a scale for which $v = R$, and multiply them by $\frac{w}{g}$ to get the values of the ordinates of the acceleration curve E E in fig. 42.

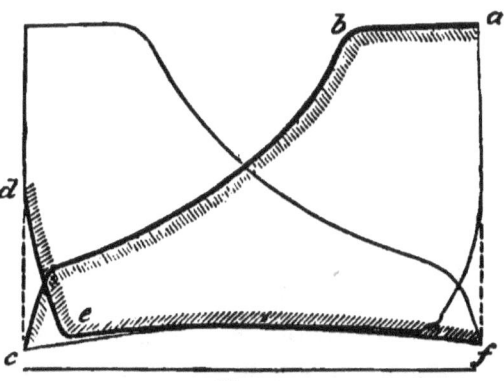

Fig. 46.

47. *Determination of curve of crank-pin effort or twisting moment when the indicator diagram of the engine is given.*—Let fig. 46 represent a pair of indicator diagrams as ordinarily taken. For the forward stroke *a b c* is the forward

pressure in the back end of the cylinder and def the simultaneous back pressure in the forward cylinder end. The vertical intercept between these lines is the effective forward pressure at any instant. These vertical intercepts have been set off from the horizontal ae, fig. 47, to form the figure $ahkle$, which represents the effective forward thrust, due to steam pressure, at the crosshead. It will be seen that at the crank

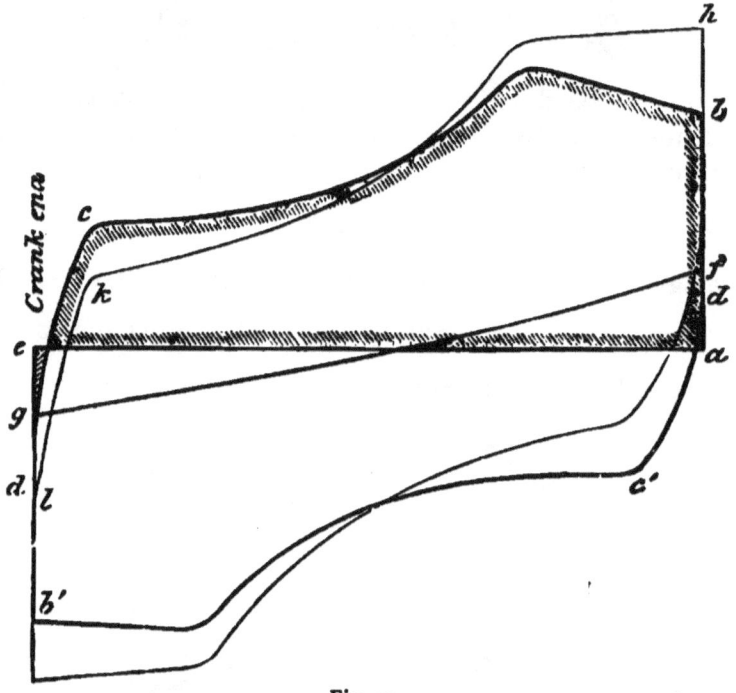

Fig. 47

end, for a very short part of the stroke, the pressure is negative. The diagram has now to be corrected for the inertia of the reciprocating parts.

In fig. 47, fg is the curve the vertical ordinate of which is the force due to inertia, reckoned like the steam pressure per unit of area of piston. During a little less than half the forward stroke the force due to inertia is negative, but is set off upwards so as to be deducted from the steam pressure. Similarly, during the rest of the forward stroke the force due to inertia is positive, but is set off downwards so as to be added to the steam pressure. Finally, then, the vertical intercept between $a\,h\,k\,l$ and $f g$ is the effective forward pressure, including the effect of inertia. For convenience, these vertical intercepts have been set off to form the shaded figure $a\,b\,c\,d$, the ordinates of which are the real effective forward pressures after adding or deducting the force due to inertia. Similarly $e\,b'\,c'\,d'$ is the corresponding figure for the return stroke. It is the ordinates of these figures which are to be used in the construction shown in fig. 42 in finding the real crank efforts and crank shaft twisting moments.

48. *Combination of effort curves for two or more engines.* — Suppose there are two engines at right angles. Then it is convenient to draw the piston effort curves, such as those in fig. 47, in positions at right angles, and so to obtain the polar curves for each engine in a convenient position for adding the vectors corresponding to each engine to obtain the combined polar curve. This has been done for a compound engine with cranks at right angles in fig. 48.

If the engine is a compound one, a further reduction must be made. Either the high-pressure diagram is drawn to natural scale, and for the low pressure the pressures are reduced to equivalent pressures on the high-pressure piston, or else the low-pressure diagram being drawn to natural scale, the high-pressure diagram pressures must be reduced to equivalent pressures on the low-pressure piston. If the low-pressure piston area is n times the high pressure, then either the low-pressure diagram ordinates must be increased in the ratio $n:1$ or the high pressure reduced in the ratio $1:n$. Then an ordinate on either diagram represents the piston effort in lbs. per unit of area of the piston to which the pressures are reduced. In the figure the curves A with a simple line are the steam pressure and inertia curves as

78   *Machine Design*

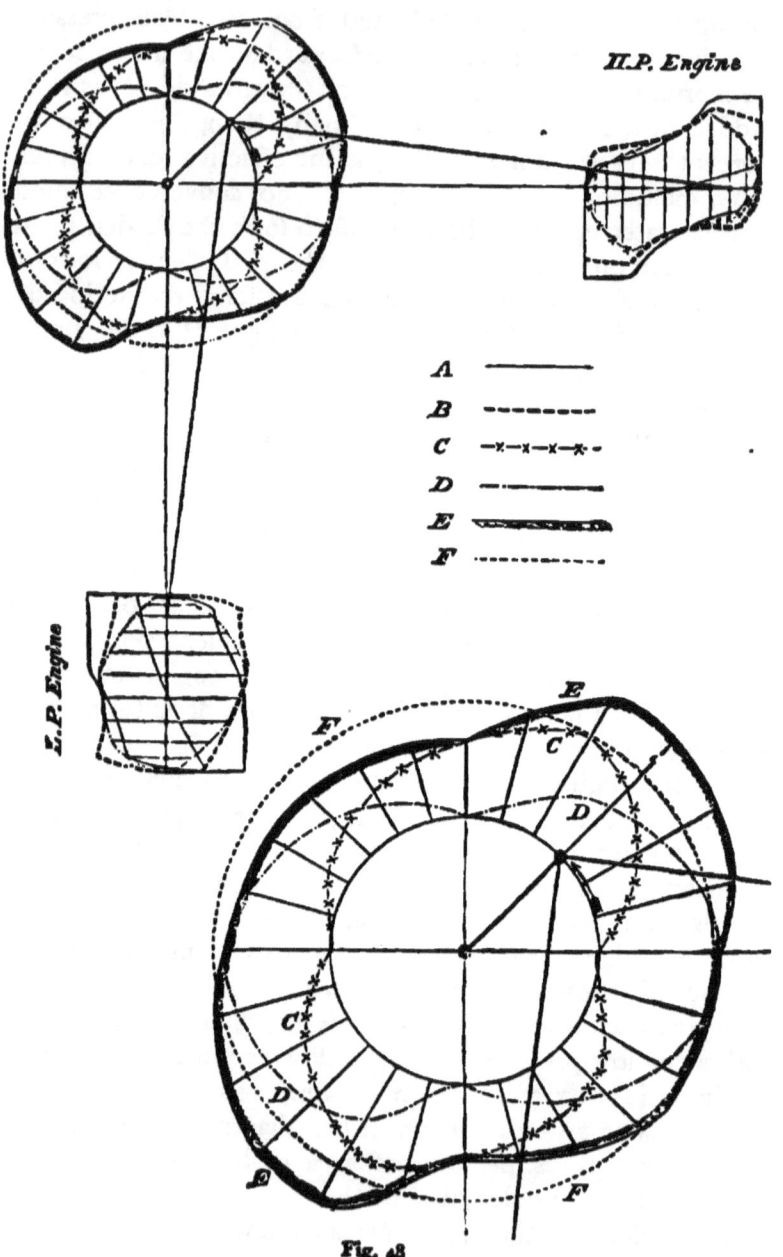

Fig. 48

in fig. 47. The resultant piston efforts are given by the dotted curves B. The curves C are the true piston effort curves corrected for inertia for the H.P. cylinder and D those for the L.P. cylinder. Adding the ordinates of these in the polar diagram, the curve of total effort E due to both engines is obtained. The dotted circle F is a curve of mean effort.

49. *Direct influence of the weight of reciprocating parts in vertical engines.*—In vertical engines the weight of the piston, crosshead, &c. acts in the same direction as the steam pressure in the downward stroke and in the opposite direction to the steam pressure in the upward stroke. If $w_1 + w_2$ is the weight of the reciprocating parts reckoned per unit of piston area, then the effective piston effort is increased by $w_1 + w_2$ throughout the down stroke and diminished by the same amount throughout the up stroke. The correction for this action of the weight is most conveniently made by shifting the base line of the piston effort diagram a distance $w_1 + w_2$ so as to increase the area of the top and diminish the area of the bottom diagram before drawing the inertia curve.

The total weight $w_1 + w_2$ per sq. in. of piston is about 2 to 4 lbs. in most engines, occasionally reaching 6 lbs.

50. *Advantages and disadvantages of the inertia of the reciprocating parts.*—In every reciprocating engine the resultant force at the crank pin changes from a push to a pull and back again to a push in every revolution. Now, as there must always be some slack in the crank pin and crosshead brasses, there is a liability to a knock of more or less violence if this change occurs suddenly or at much velocity. Probably the liability to an injurious knock is least if the change occurs exactly at the dead points, where the motion in the direction of the line of stroke is limited to the slack of the brasses. A knock may be produced even at this point, if the

initial pressure of the steam is very great, and compression is supposed to be useful in preventing a knock of this kind. Probably, however, a serious knock is generally produced in another way. If there is much compression and if the inertia of the reciprocating parts is great, the direction of the effort at the crank pin may not change till some time after the beginning of the stroke; or, in other words, for a sensible part of the stroke the crank-pin effort acts opposite to the piston effort. This occurs if the inertia line, gf, fig. 47, rises above the piston-effort line $h\,c\,d$, at the beginning of the stroke. Then a knock occurs from the reciprocating parts catching up with the crank pin. The heavier the reciprocating parts, the greater the speed of the engine and the greater the compression, the more likely is it that there will be a negative crank effort for part of the beginning of the stroke. Hence, taking it for granted that the weight of the reciprocating parts cannot practically be much modified, the arising of this negative crank effort fixes a limit of speed at which the engine can be run quietly. In order to get rid of the tendency to knock, some high-speed engines have been built with hollow piston rods, and in other cases single-acting engines have been adopted, so arranged that in all conditions there is a thrust between the connecting rod and crank.

On the other hand, the action of the inertia of the reciprocating masses is advantageous in equalising the crank, pin effort in engines having much expansion. The work expended in accelerating the reciprocating masses in the first half of the stroke when the steam pressure is high is given back in the second half of the stroke when the steam pressure has fallen from expansion. By suitably choosing the weight of the reciprocating masses for any given ratio of expansion a very uniform crank-pin effort can be secured. Generally, however, it is more convenient to depend on a fly wheel to equalise the twisting moment than to alter the weight of the piston or crosshead.

Hand Levers and Winch Handles

51. Fig. 49 shows an ordinary straight lever for working machinery by hand. The part grasped by the hand may be $1\frac{1}{4}$

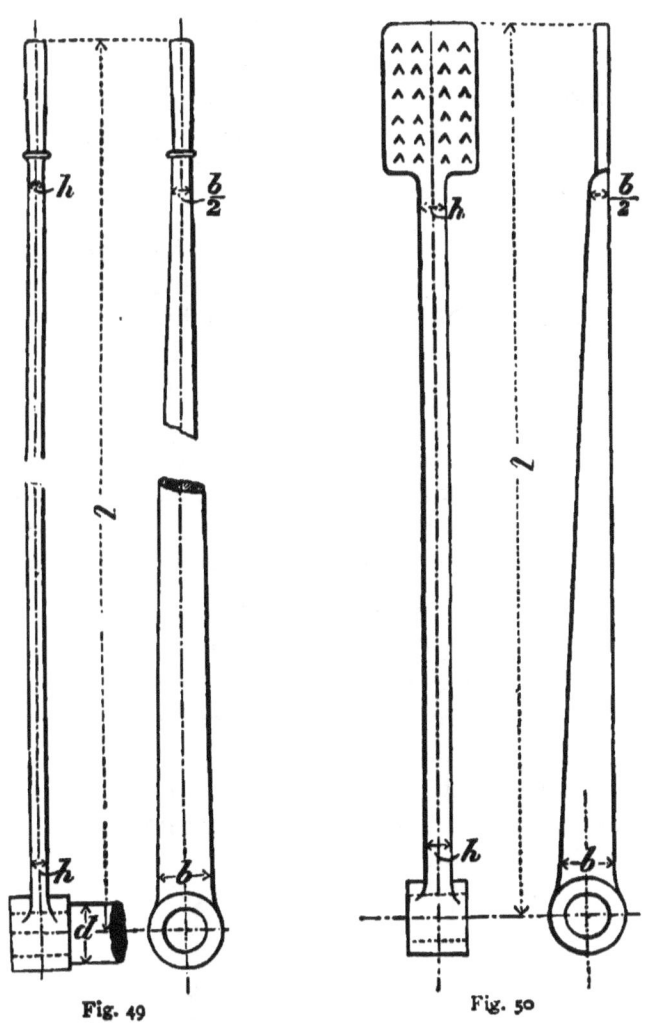

Fig. 49   Fig. 50

inch in greatest and 1 inch in smallest diameter, and 5 inches long. Let P be the force exerted at the handle, and l the length of the lever. Then Pl (nearly) is the greatest bending moment on the arm. Let b be the width and h the thickness of the arm at its largest part. Then, § 28 (Part I.),

$$\tfrac{1}{6} b^2 h f = Pl$$
$$b^2 h = \frac{6\,Pl}{f}$$

Let the greatest force, P, exerted by a man be taken at 84 lbs.; and let $f = 9,000$ lbs. per sq. in. for wrought iron. Then,

$$b^2 h = \tfrac{1}{18} l \text{ nearly} \qquad . \qquad . \qquad . \quad (6)$$

$h = \tfrac{3}{4}$ inch, $b = 0.27\sqrt{l}$. If $h = \tfrac{1}{2} b$, then $b = 0.5 \sqrt[3]{l}$ and $h = 0.25 \sqrt[3]{l}$ nearly. If the flat part of the lever is of uniform thickness, its least width should be half its greatest width, the case corresponding with Case I. Table VII. (Part I.). Let $d =$ diameter of shaft on which the lever is keyed; $n =$ distance from centre of lever to centre of nearest bearing of shaft. Then the shaft is subjected to a twisting moment Pl and a bending moment Pn, and its strength is determined by the rules in I. § 44 and § 137. The equivalent bending moment is $P(0.7n + 0.48l)$ nearly. Hence,

$$d = 0.0947 \sqrt[3]{\{P(1.4n + 0.96l)\}}$$
$$= 0.42 \sqrt[3]{(1.4n + 0.96l)} \qquad . \qquad . \qquad . \quad (7)$$

The part in the eye of the lever may have a diameter $= 0.42 \sqrt[3]{l}$. The eye of the lever may have a thickness $= 0.3 d$ and a length $= 1$ to $1\tfrac{1}{4} d$.

Fig. 50 shows a foot lever. The foot plate is about 8 ins. by 5 ins., and $\tfrac{5}{8}$ in. thick. In designing this lever P may be taken at 180 lbs. Then,

$$\left. \begin{array}{l} b^2 h = \tfrac{1}{8} l \\ d = .54 \sqrt[3]{(1.4n + 0.96\,l)} \end{array} \right\} \qquad . \quad . \quad (8)$$

Fig. 51 shows a winch handle or cranked lever. When this is intended to resist the full force of one man, P may be taken at 84 lbs., and if worked by two men, $P = 168$ lbs. The mean effort per man in continuous work is only 15 to

30 lbs. The radius r is usually 16 or 17 ins., and the height of the shaft from the ground may be 3 ft. to 3 ft. 3 ins. The length of the handle l may be 10 or 12 ins. for one man and 20 ins. for two men. The pressure on the handle may be taken to act at ⅔rds of the length. The greatest bending moment at the handle is $\tfrac{2}{3} P l$. Then its diameter should not be less than

$$d = 0{\cdot}0947 \sqrt[3]{\tfrac{2}{3} P l} = 0{\cdot}1042 \sqrt[3]{P l} \quad . \quad (9)$$

or say $1\tfrac{1}{16}$ inch for one man and $1\tfrac{1}{2}$ inch for two men. The journal of the shaft is subjected to a twisting moment $P r$,

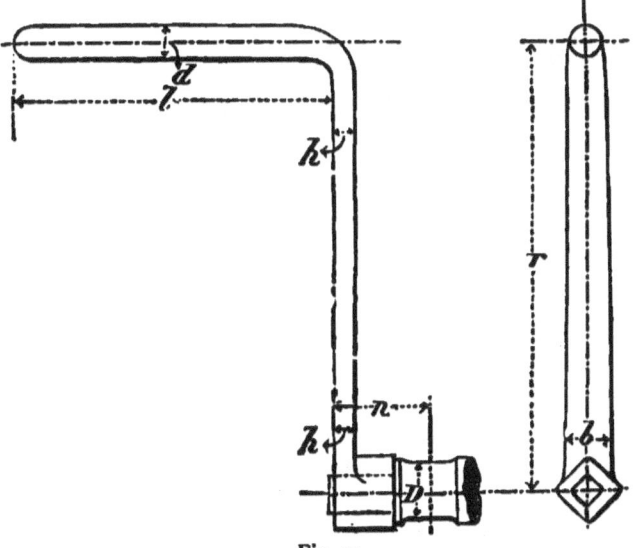

Fig. 51

and a bending moment $P(\tfrac{2}{3} l + n)$. The equivalent bending moment (I. § 44) is $P(0{\cdot}6 l + 0{\cdot}9 n + 0{\cdot}4 r)$ nearly. Then,

$$\begin{aligned}
D &= 0{\cdot}0947 \sqrt[3]{\{P(1{\cdot}2 l + 1{\cdot}8 n + 0{\cdot}8 r)\}} \\
&= 0{\cdot}42 \sqrt[3]{(1{\cdot}2 l + 1{\cdot}8 n + 0{\cdot}8 r)} \text{ for one man} \\
&= 0{\cdot}54 \sqrt[3]{(1{\cdot}2 l + 1{\cdot}8 n + 0{\cdot}8 r)} \text{ for two men}
\end{aligned} \right\} (10)$$

For the part in the eye of the crank the term $1{\cdot}8 n$ may be omitted. The greatest bending moment on the arm is $P r$, and the twisting moment $\frac{2}{3} P l$ nearly. Hence the equivalent bending moment is $P(0.9\, r + 0.27\, l)$. If b is the breadth and h the width of the arm at the larger end,

$$b^2 h = \frac{6 P}{f}(0.9\, r + 0.27\, l)$$
$$= 0.056\,(0.9\, r + 0.27\, l) \text{ for one man}$$
$$= 0.112\,(0.9\, r + 0.27\, l) \text{ for two men} \quad (11)$$

Either b or h may be selected and the other obtained from the formula. If the arm is of uniform thickness, its least breadth should not be less than $\frac{1}{2} b$, or less than $2\sqrt{\frac{P l}{f h}}$ or $0.19\sqrt{\frac{l}{h}}$ for one man, and $0.27\sqrt{\frac{l}{h}}$ for two men.

Thickness of eye of crank, $0.3\, D$; length of eye, $1\frac{1}{4}\, D$.

ENGINE CRANKS

52. Engine cranks are of cast- or wrought-iron.—A single crank consists of a nave bored to receive the crank shaft, an arm, and a crank pin. If the crank pin is a separate piece, it is fitted into an eye formed at the small end of the crank. Disc cranks have plain circular discs, instead of the ordinary crank arm, and they have the advantage of being nearly balanced with respect to the crank shaft. A double crank is used when the crank pin cannot be placed at the end of the crank shaft. An eccentric is a crank of peculiar form. It is essentially a crank, with a crank pin, the radius of which is greater than the sum of the crank and crank-shaft radii.

53. *General case. Straining action on crank arm.*—Let fig. 52 represent a crank in any position, and let P be the total pressure on the crank pin. Resolve P into a tangential component T, and a radial component N. Let ab be any section of the arm at a distance r from the centre of crank pin, and let m be the distance between centre lines of crank pin and crank arm. Then the straining actions at ab which require to be considered are :—

(*a*) A direct pressure (or tension) equal to N.
(*b*) A bending moment N m in the plane of the arrow B.
(*c*) A bending moment T r in the plane of the arrow A.
(*d*) A twisting moment T m.

To take into account all these straining actions in several positions of the crank would be laborious. Generally it is sufficient to estimate the strength of the crank in two positions, when the crank is at the dead point and when the crank and connecting rod are at right angles. In the former case, T vanishes and N becomes equal to the greatest piston pressure, or to $N'' = P'' - \dfrac{W v^2}{g R}$, which will also for simplicity be denoted by N simply. Here P'' is the initial piston load, W the weight of reciprocating parts, V the velocity of crank pin and R its radius. In the latter case N vanishes and T is equal to 1·02 or 1·03 time the piston pressure.

54. *Strength of the crank.*—Let N be the radial pressure when the crank is at the dead point, and T the tangential pressure when the crank and connecting rod are at right angles. Let, further,

d, l = diameter and length of crank pin.
D, L = diameter and length of crank-shaft journal.
d', l', t' = internal diameter, length and thickness of small eye of crank.
$d'' l'' t''$ = internal diameter, length, and thickness of large eye of crank.
h, b = thickness and width of arm at any section ab; the same letters with one accent referring to the section of the arm supposed produced to the centre of small eye, and with two accents the section produced to the centre of large eye.
R = crank radius.

m = distance from centre line of crank pin to centre line of crank arm.

n = distance from centre line of crank pin to centre line of crank-shaft journal.

The crank pin and crank-shaft journal are first designed by the rules in I. Chapter VII. For the section of the crank

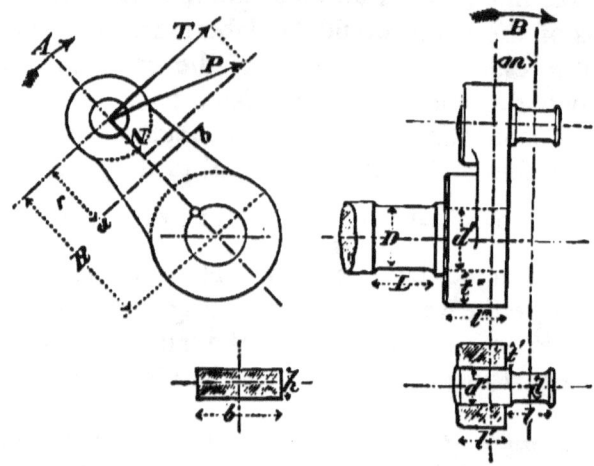

Fig. 52

arm we have, when the crank is at the dead point, a direct tension or pressure N and a bending moment N m. Then the greatest stress is (L § 43),

$$f = N\left(\frac{1}{hb} + \frac{6m}{bh^2}\right)$$

$$b = \frac{N}{f}\left(\frac{1}{h} + \frac{6m}{h^2}\right) \quad . \quad . \quad . \quad (12)$$

or, $\quad h = 2 \cdot 65 \sqrt{\dfrac{Nm}{fb}}$ nearly . . (13)

If this straining action only were considered, the crank arm would be of uniform section throughout. Hence this equation is chiefly useful for determining the breadth and thickness of the arm at the small end, where the straining action due to the force T is least important.

When the crank and connecting rod are at right angles, there is a bending moment Tr, and a twisting moment Tm, Combining these, the equivalent bending moment is (I. § 44)

$$= 0\cdot 91\, Tr + 0\cdot 41\, Tm \text{ nearly.}$$

The bending is parallel to the plane of rotation, so that the modulus of the section is $\tfrac{1}{6} b^2 h$. (I. Table V.) Then

$$\tfrac{1}{6} b^2 h f = 0\cdot 91\, Tr + 0\cdot 41\, Tm$$

$$b = \sqrt{\left\{\frac{6\,T}{fh}(0\cdot 91\, r + 0\cdot 41\, m)\right\}} \quad . \quad (14)$$

$$\text{or, } h = \frac{6\,T}{f b^2}(0\cdot 91\, r + 0\cdot 41\, m). \quad . \quad . \quad (15)$$

Select a value for h or b. Then the greater of the values given by equations 12 and 14, or 13 and 15, is the proper value for the remaining dimension. It will often be sufficient to use equation 12 or 13 to determine b' or h'; equation 14 or 15 to determine b'' or h''; and the sides of the crank arm may be drawn as planes.

When the crank is of cast iron the arm may be trough-shaped (fig. 54), and is then somewhat lighter than when it is rectangular. Let b and h be the width and thickness of a rectangular arm, and let $b_1\, h_1$ and $b_2\, h_2$ be the dimensions of an arm of trough-section of equivalent strength. Then, if flexure is in the plane of rotation,

$$b^2 h = \frac{b_1{}^3 h_1 - b_2{}^3 h_2}{b_1}$$

Let $\quad b_1 = b$

$$b^3 h = b_1{}^3 h_1 - b_2{}^3 h_2$$

Let $\quad b_2 = x\, b$

$\quad h_2 = y\, h_1$

Then $\quad h_1 = \dfrac{h}{1 - x^3 y} = c\, h. \quad . \quad (16)$

where c has the following values :—

$y=$	0·6	$x=$ ·65	·7	·75
·6	1·15	1·20	1·26	1·34
·65	1·16	1·22	1·29	1·38
·7	1·18	1·24	1·32	1·42
·75	1·19	1·26	1·35	1·46

At the section at the centre of the small eye of the crank the flexure is at right angles to the plane of rotation, and the

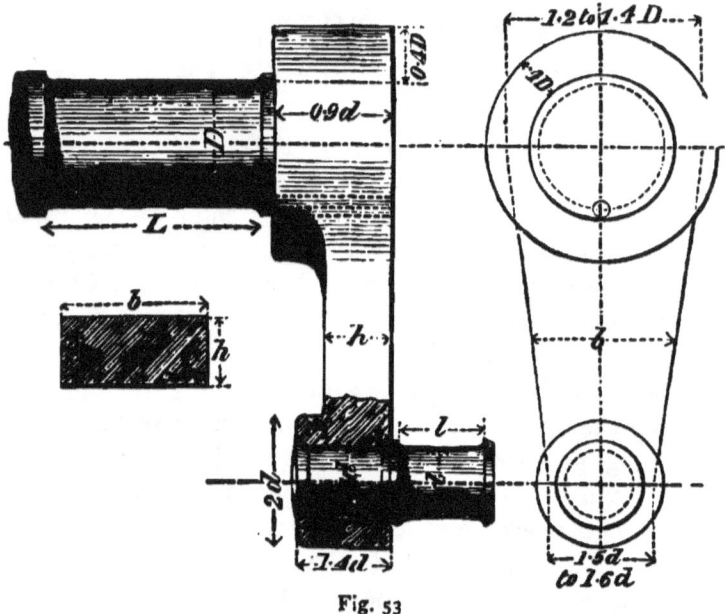

Fig. 53

feathers strengthen the section very little. Hence the section there may remain unchanged, the feathers being allowed to diminish towards that end.

When a crank has a T-form of section, it is but little strengthened by the feather, but it is more easily cast.

55. *Proportions of cranks.*—The crank is shrunk on to the crank shaft, and the crank pin is also fixed in the same way and riveted cold. The key in the crank shaft may have a breadth $= \frac{1}{3}$ D, and a thickness $\frac{1}{6}$ D, for small cranks, and $\frac{1}{4}$ D and $\frac{1}{8}$ D for large cranks.

Fig. 53 shows a wrought-iron crank with a section of the arm. The arm is sometimes tapered and the back face of the arm is then rounded, so that it forms, in fact, part of a slightly conical surface, turned in the lathe. Fig. 54 shows

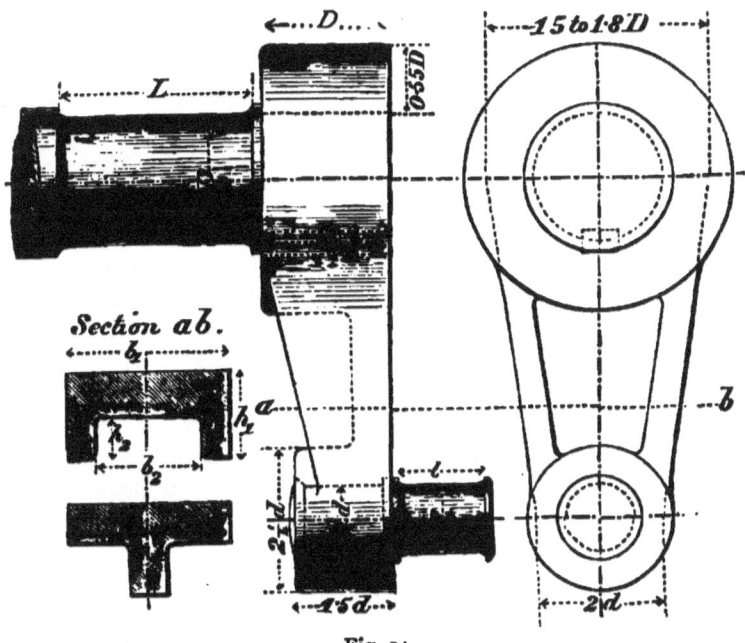

Fig. 54

a cast-iron crank, with sections of arms both trough-shaped and T-shaped.

In quick-running engines it is desirable to balance as directly as possible the weight of the crank and connecting rod. The crank then takes a disc form, as shown in fig. 55. By hollowing the part of the disc on the crank side and leaving the opposite side full, a surplus weight is obtained which balances the crank pin and connecting-rod end.

The question of the amount of balance weight required

involves some difficulty, because parts of the weights attached to the crank pin reciprocate without rotating, and part rotate with the crank pin. Let w_1 be the weight of the balance weight, and ρ the radius to its centre of gravity; w_2 the weight of the crank pin and half the weight of the connecting rod, which may be taken as rotating with the crank pin at radius r; w_3 the weight of the piston, piston rod, crosshead, and the other half of the weight of the connecting rod. Then for balance of the forces at right angles to the line of stroke

$$w_1 \rho = w_2 r.$$

But for balance of the forces parallel to the line of stroke

$$w_1 \rho = (w_2 + w_3) r.$$

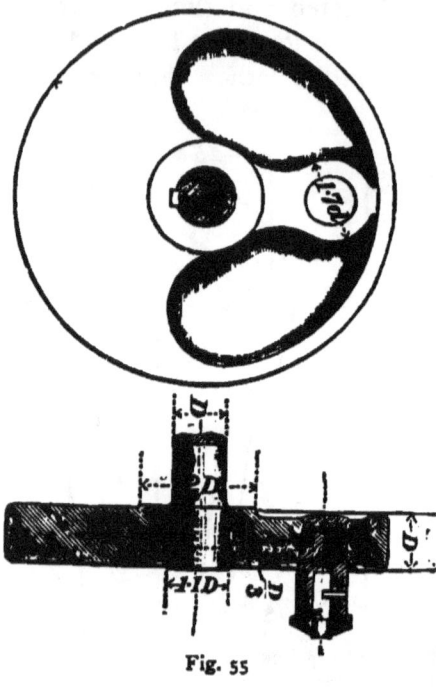

Fig. 55

Hence the axial and normal disturbing forces cannot both be balanced. If the balance weight is sufficient for the axial forces, it over corrects the normal forces and introduces a new unbalanced force perpendicular to the line of stroke. In ordinary practice for vertical engines, in which the forces at right angles to the line of stroke are most injurious,

$$w_1 = w_2 \frac{r}{\rho}.$$

But for horizontal engines, where the horizontal forces are most injurious,

$$w_1 = \tfrac{2}{3}(w_2 + w_3)\tfrac{r}{\rho} \text{ to } \tfrac{3}{4}(w_2 + w_3)\tfrac{r}{\rho}.$$

In locomotives, when the balance weight is made as large as in the latter case, the vertical unbalanced forces are considerable, and act dangerously in tending to throw the engine off the rails, or, at all events, tend to damage the wheels and rails. Consequently for locomotives,

$$w_1 = w_2\tfrac{r}{\rho}.$$

56. *Built-up steel cranks.*—The difficulty of forging large double cranks has led to the use of built-up cranks like that

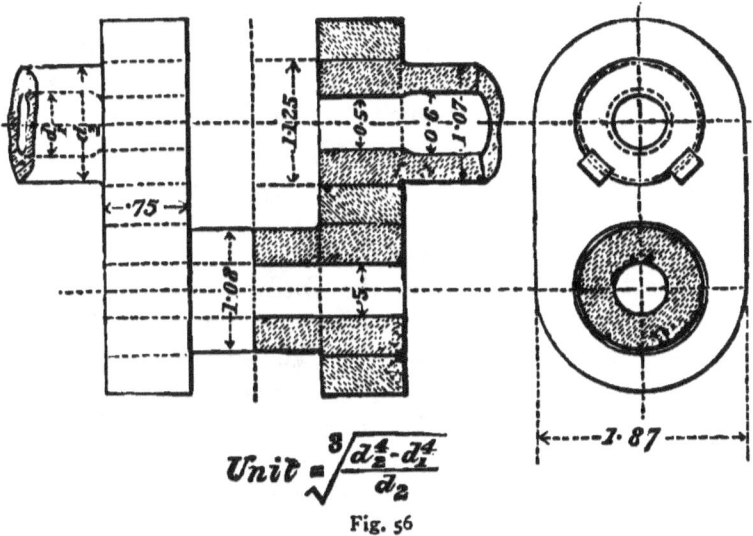

Fig. 56

shown in fig. 56, which shows the cranks used in the 'City of Rome' s.s.—A double-collared hollow steel shaft, formed as described in I. Chapter VII. § 152, is cut in half to form the single-collared pieces. The crank cheeks or webs are first forged solid in the form of slabs, and then a small hole is bored at each end, and enlarged by being forged on a mandril placed on suitable supports ; thus insuring that the metal is thoroughly worked in the most important part. The cheeks are afterwards shrunk and keyed on the cut ends of the half lengths of shaft. The hollow crank pin is drawn to length by forging, and is shrunk in but not keyed into the cheeks.

ECCENTRICS

57. An eccentric is a modified crank, chiefly employed to drive the slide valve of steam engines. It is really a crank and connecting rod, with a crank pin enlarged, so as to include the crank shaft within its section, the radius of the eccentric being greater than the sum of the crank and crank shaft radii. The eccentric consists of a sheave, which is virtually a crank pin, and a strap and rod which is virtually equivalent to a connecting rod. The sheave is most commonly of cast iron, and is often cast in two parts connected by bolts. In very hard-worked eccentrics the sheave may be of wrought iron, case-hardened. When the sheave is in two parts, the smaller may be of wrought and the larger of cast iron. The strap is in two parts, and is prevented from slipping sideways by a flange or flanges, or it has internally a spherical surface fitting on the sheave. The strap is of brass, of cast iron, or of wrought iron lined with brass or with white metal. It is doubtful if any eccentric strap wears as well as one of cast iron. The friction of the eccentric is much greater than that of a crank, and it is therefore not used where ordinary cranks can be applied.

The distance between the centres of the crank shaft and eccentric sheave is termed the 'eccentricity,' the 'radius,' or the 'half stroke' of the eccentric. Let this be denoted by r, and let d be the diameter of the shaft on which the eccentric is fixed. Then the least diameter suitable for the eccentric sheave is about

$$= D = 1{\cdot}2\,d + 2\,r + \tfrac{3}{4}.$$

58. *Width of bearing surface of eccentric sheave.*—The sheave is properly a journal of much enlarged diameter. In such a case, as the velocity of rubbing is much greater than for ordinary journals, the pressure on the bearing surface must be less. The rule which gives the most satisfactory agreement with practice is that in I. § 124, equation 10.

Let P be the total thrust exerted by the eccentric, usually to overcome the friction of a slide valve; let b be the width of eccentric sheave and N the number of revolutions per minute. Then

$$b = \frac{PN}{60,000},$$

a value agreeing fairly well with that for ordinary crank pins.

59. *Friction of the slide valve.*—Let a be the area of the back of the valve subjected to the steam pressure, and p the steam pressure in lbs. per sq. in., reckoned above atmospheric pressure in the case of non-condensing engines, and above zero in the case of condensing engines.

Then the frictional resistance is ordinarily taken to be

$$P = \mu p a,$$

where μ is about 0·10 for smooth surfaces, such as slide-valve surfaces, not well lubricated.

In the following figures the unit taken is

$$k = \frac{PN}{25,000} = \frac{paN}{170,000}.$$

Some interesting experiments have been made by Mr. Aspinall on the friction of locomotive slide valves ('Proc. Inst. Civil Engineers,' vol. xcv. p. 176). He takes the load on the valve to be the valve-chest pressure on the back of the valve, less the cylinder pressure on one steam port and less the back pressure in the exhaust passage. The pressures were determined from indicator diagrams. The measurement of the resistance to the motion of the valve at mid-stroke was also measured. The mean coefficient of friction was 0·068, 0·054 and 0·051 for three different valves. The relief of pressure on the area of the steam port and exhaust

94   *Machine Design*

passage was not a large fraction of the total pressure on the back of the valve, so that probably the real friction is about one-half of that found by the expression above. However, as slide valves are not always in good condition, and the eccentric must be calculated for the worst case, the value of P above may be taken in estimating the width of eccentric sheave.

60. *Radius of eccentric.*—Let w be the greatest width of

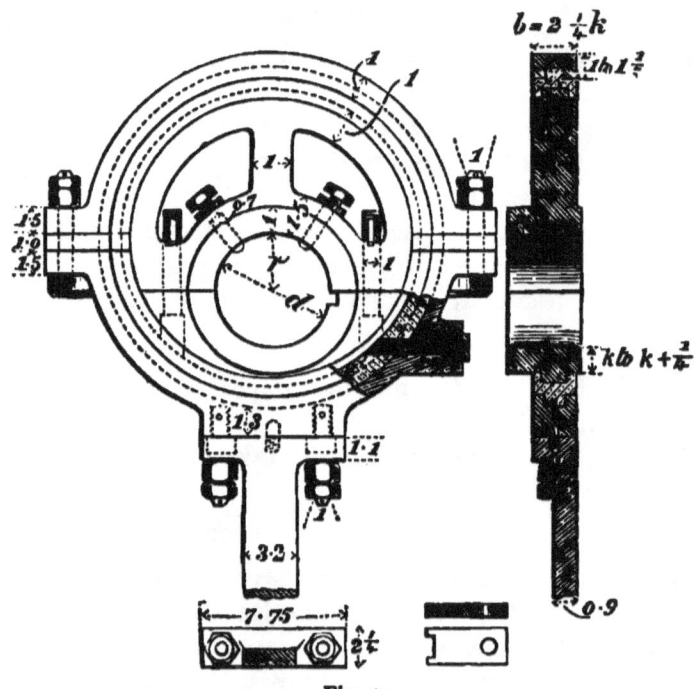

Fig. 57

port opened to steam ; l, the lap of the valve ; r, the radius of the eccentric,

$$r = w + l,$$

w is in some cases the whole width of the steam port, but in quick-running engines the opening to steam is less than the opening to exhaust. This is secured by making

w about ⅔rds of the width of the port. The external lap l may vary from ⅛th of the width of the port to the whole width of the port, according to the amount of expansion required.

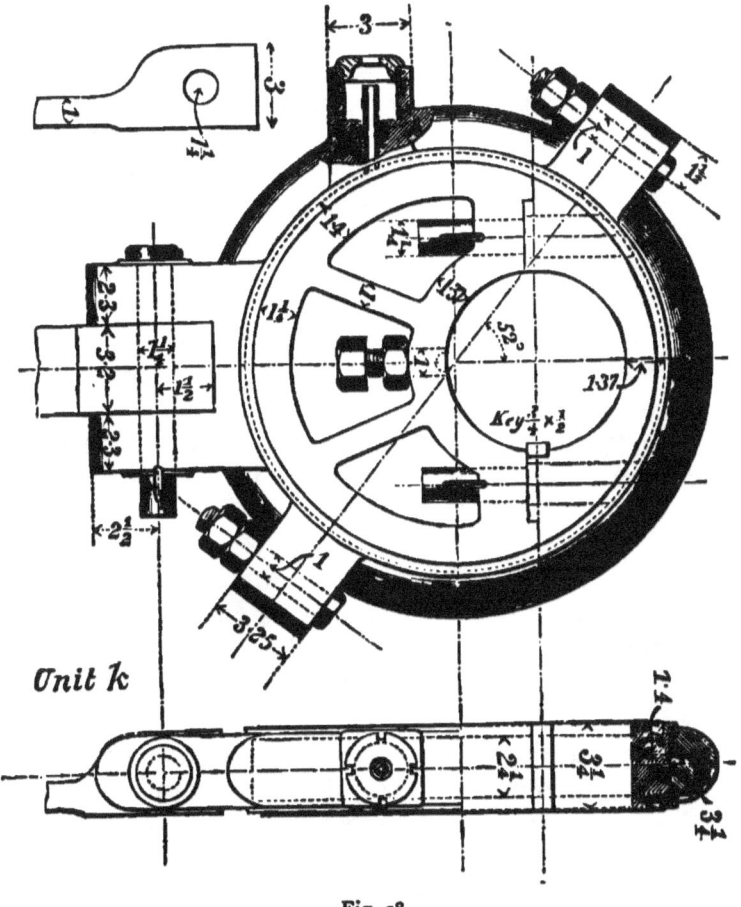

Fig. 58

The proportions of slide valves will be given in a later chapter.

61. *Proportions of sheave.* — The width b of the bearing surface of the sheave may be taken equal to $2k$, or $2\tfrac{1}{4}k$.

The diameter of the bolts connecting the two parts of the sheave may be $0.85\,k$ to k, and the cotter in these bolts may have a width equal to their diameter and a thickness equal to $\frac{1}{4}$ of their diameter. The set screws may be $0.7\,k$ diameter, or if there is only one it may be k in diameter. If the sheave has flanges to retain the strap, their projection may be $0.4\,k$, and their thickness $0.3\,k$. The smaller part of the sheave, when it is in two parts, is generally of wrought iron in small eccentrics. The least thickness of this part, where the shaft comes near the edge of the sheave, should be at least k for wrought iron and $1\frac{1}{4}\,k$ for cast iron.

62. *Proportions of strap*.—The strap thickness varies very much. For wrought iron it may be from $.5\,k$ in large to k in small eccentrics. For gun-metal it should be from $.625\,k$ to $1\frac{1}{4}\,k$. For cast iron from $.75\,k$ to $1.5\,k$. The brass lining may be about $\frac{1}{8}$th the strap thickness in large eccentrics, and in other cases its thickness may be

$$\frac{D}{40} + \frac{1}{8}.$$

When the strap is recessed to fit a projection on the sheave, the depth of the recess may be $\frac{2}{3}$ths of the thickness of the brass, and its width $0.3\,b$. The corresponding recess in the brass may be of the same depth, and its width $b - \frac{5}{8}$ to $b - \frac{7}{8}$.

With eccentric straps made in the ordinary manner there is a tendency to open at the joint slightly, causing cutting and heating. This may be avoided by making the strap of the form shown in fig. 59.[1] The bolts are long and brought as close as possible to the sheave. The bolts bear in the strap only at the middle and ends, and are turned down to the diameter at bottom of screw thread in intermediate parts. This gives a little elasticity to the bolts and prevents

---

[1] The eccentric shown in fig. 59 was designed by Mr. Druitt Halpin.

Linkwork—Cranks and Eccentrics

the fatigue of the metal at the last screw thread. An oil box can be formed in each half of the strap.

Proportions of eccentric rod.—The eccentric rod is very commonly attached to the eccentric strap by a T-end, and has at the other an eye to receive the pin of the valve rod. At its smaller or eye end it may have a width of $1.8\,k$ and a thickness $0.55\,k$. It tapers in width about $\frac{1}{2}$ inch per foot

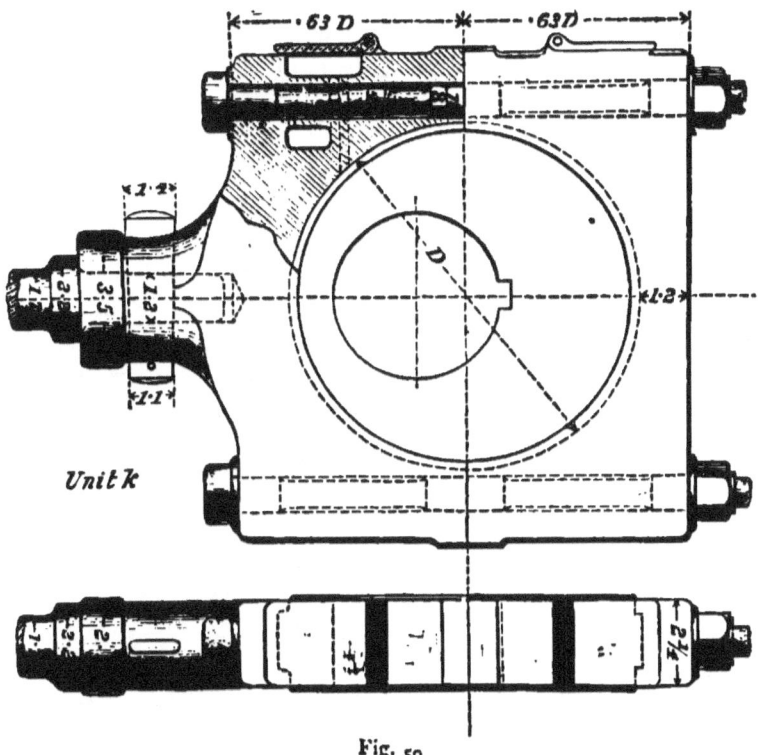

Fig. 50

of length to the T-end, the thickness being constant. The bolts in the T-end may be of the same size as the strap bolts.

Fig. 58 shows a link-motion eccentric for a locomotive having both sheave and strap of cast iron, and so arranged as to be easily taken apart.

In both fig. 57 and fig. 58 the eccentric sheave is shown divided into two parts for convenience of fixing. When the eccentric can be put on from the end of the shaft this is not necessary.

63. *Friction of eccentric.*—Let R be the radius of the eccentric sheave, in ins., P the resistance of the slide valve or other part moved by the eccentric, in lbs., N the number of rotations per minute, μ the coefficient of friction. Then the frictional resistance at the surface of the sheave is about μ P lbs., and the work expended in friction is

$$\frac{2 \mu P \cdot \pi R N}{12 \times 60} \text{ ft. lbs. per sec.}$$

Or, putting $\mu = 0.06$,

$$0.00052 \text{ P R N ft. lbs. per sec.}$$

This is so large that in some cases it amounts to 20 or even 25 per cent. of the whole work transmitted by the eccentric. Cast-iron sheaves and straps appear on the whole to wear less than any others. Hence, probably, also they work with the least friction.

CHAPTER IV

CONNECTING RODS

64. Connecting rods are the pieces which connect a rotating crank with a reciprocating piece, such as a piston or pump plunger. A link connecting two cranks is generally termed a coupling rod. In the older steam engines the connecting rod was often of cast iron, and frequently it had a cross-shaped section in order to get sufficient stiffness without too much increasing the section. Now most connecting rods are of wrought iron or steel. For engines of moderate speed the connecting rod is generally of circular section, tapered from the centre towards the ends or from the small end towards the big end, with a view of lightening the rod without much diminishing its resistance to bending. In high-speed engines, as, for instance, in locomotives, greater resistance to bending in the plane of oscillation is obtained by making the rod of rectangular or approximately rectangular section. Sometimes even an **I**-shaped section is adopted in locomotive coupling and connecting rods, part of the sides of a rectangular rod being cut away by milling to leave an **I**-shaped section.

The ends of a connecting rod are fitted with brass steps and adjusting arrangements for neutralising wear, and these are fitted to the crank pin at one end and the crosshead pin at the other. If, in consequence of adjusting the brasses, the connecting rod is altered in length, the clearance at the two ends of the cylinder is altered. To prevent this, it is desirable to make the adjustments for wear at the two ends of the rod in such a way that tightening up the brasses at one end lengthens the rod and tightening at the other shortens it.

In consequence of the varying obliquity of the connecting rod, the position of the piston is not the same for corresponding crank angles in the forward and return strokes. With a very short connecting rod this has a prejudicial effect on the distribution of steam, and at the same time the pressure on the guides becomes excessive. Hence, usually in steam engines the connecting rod is 4 to 6 cranks in length. For engines at high speed a long connecting rod is specially desirable. In the Westinghouse single-acting engine the line of stroke passes to one side of the centre of the crank shaft. The effect of this is to diminish the obliquity of the connecting rod (or virtually to lengthen it) during the working stroke and to increase the obliquity in the non-working or return stroke.

65. *Straining forces acting on the connecting rod.*—Let p be the greatest effective steam pressure on the piston, reckoned from atmospheric pressure if the engine is non-condensing, or from the condenser pressure if the engine is condensing. For compound-engine cylinders p is to be taken as the greatest difference of pressure on the two sides of the piston. Let D be the diameter of the cylinder. Then

$$P = \frac{\pi}{4} D^2 p \quad . \quad . \quad . \quad . \quad (1)$$

is the effective piston load which is transmitted to the connecting rod. If the engine is starting slowly, so that the inertia forces are negligible, P is actually the thrust or pull on the connecting rod, and this must in any case be provided for. But under other circumstances the force acting on the connecting rod may be very different, some causes tending to diminish and others to increase it.

(1) For the position at the beginning of the stroke the thrust or pull on the connecting rod is diminished by the

translational inertia of the parts between it and the piston. In normal working, the initial thrust or pull would be

$$P - \frac{w}{g} \cdot \frac{v^2}{R} \text{ nearly}$$

where W is the weight of the piston and crosshead, v the velocity of crank pin, and R the radius of crank. In this position also there are no transverse forces due to the oscillation of the connecting rod.

(2) At the end of the stroke the pull or thrust is

$$P_1 + \frac{w}{g} \cdot \frac{v^2}{R}$$

where P_1 is the piston load, now diminished by the effect of expansion. Generally this will be less than the pull or thrust at the beginning of the stroke. But in some cases a serious increase may be caused by the presence of water in the cylinder, and this it is not possible to estimate.

(3) At mid-revolution the obliquity of the rod increases the thrust or pull due to the piston by about 3 to 5 per cent. —a quantity useless to take into account, seeing how rough the estimate of the straining force must in any case be. The piston load P_2 will generally be less than P, in consequence of expansion and wiredrawing. On the other hand, in this position, while the inertia forces in the line of stroke have vanished, the transverse bending forces due to the inertia of the rod itself are at their maximum. Putting A for the area of section, z the modulus of section of the rod, and M for the bending moment due to the oscillation of the rod, the stress will be

$$\frac{P_2}{A} + \frac{M}{Z}.$$

In slow-running engines this will generally be a less stress than that at the beginning of the stroke, but in fast-running engines it is almost certain to be greater.

We shall assume, first, that connecting rods have to be designed for a thrust or pull m P where m is a coefficient generally greater than unity, which allows for the inertia forces and other straining actions, which modify the stress due to the initial piston load P. m will have to be determined by an examination of actual cases.

66. Resistance of connecting rod to tension.—The minimum section of the rod must be sufficient to resist the tension. Let d_{min} be the least diameter of a rod of circular section; f the safe working stress. Then

$$\frac{\pi}{4} d^2_{min} f = m P = m \frac{\pi}{4} D^2 p$$

$$d_{min} = \sqrt{\frac{m}{f}} \times D \sqrt{p} \qquad . \qquad . \qquad (2)$$

From Table II., Part I., p. 43, the safe stress f might be 1,400 lbs. per sq. in. for cast iron, 5,000 lbs. for wrought iron, and 6,600 lbs. for soft steel. The actual stresses in connecting rods are rather less than this, and will be fairly represented by taking $m = 1\frac{1}{4}$ to $1\frac{1}{2}$.

$$d_{min} = 0.0299 \text{ to } 0.0327 \text{ D} \sqrt{p} \text{ for cast iron}$$
$$= 0.0158 \text{ to } 0.0173 \text{ D} \sqrt{p} \text{ for wrought iron}$$
$$= 0.0138 \text{ to } 0.0151 \text{ D} \sqrt{p} \text{ for steel}$$

If the engine is slow running it is hardly necessary to make any further calculation. The rod may be stiffened by making its greatest diameter 1.1 to 1.125 d_{min}. If the section is rectangular it is only necessary to make

$$(b h)_{min} = \frac{\pi}{4} d^2_{min}.$$

67. Stability of connecting rods treated as columns resisting a thrust.—When, as is most commonly the case, the rod is long in proportion to its diameter, its resistance to lateral buckling must be considered.

Let d be the diameter of the rod at the centre. Then for bending in the plane of oscillation it may be considered a strut hinged at the ends, and it corresponds to Case II., Table VIII., Part I., p. 80. Let P be the initial load, m a factor as above determined by comparison of actual cases, l the length of the rod between the end journals, $I = \dfrac{\pi}{64} d^4$ the moment of inertia of the section. Then, allowing a factor of safety of 6,

$$6 m P = \pi^2 \dfrac{E I}{l^2} \quad . \quad . \quad . \quad . \quad (3)$$

$$d = \sqrt[4]{\dfrac{384}{\pi^3 E}} \sqrt[4]{l^2 P} \sqrt{m} \quad . \quad . \quad (3a)$$

where $\sqrt[4]{\dfrac{384}{\pi^3 E}} = 0\cdot 0256$ for wrought iron or steel.

For locomotives it appears that m has a value sometimes as low as 2, but for stationary and marine engines it is more commonly 4 or 6. Hence

$m =$	2	3	4	5	6
$\sqrt[4]{m} =$	1·190	1·316	1·414	1·496	1·565
$\sqrt[4]{\dfrac{384}{\pi^3 E}} \sqrt[4]{m} =$	0·0304	0·0336	0·0361	0·0382	0·0400

It will be seen that the diameter only increases slowly for a great increase in the value assumed for m, and this partly explains why in practice such greatly divergent values are found in different cases.

Connecting rods of rectangular section for quick speeds.—Let b be the breadth and h the height of the section, the latter being in the plane of oscillation of the rod. Then two cases are to be considered. The rod may bend in the plane of oscillation, in which case it is to be considered as hinged at the ends, and it corresponds to Case II., Table VIII., Part I. If it bends at right angles to the plane of oscillation it approximates to the condition of a strut fixed at the ends : Case IV., Table VIII. Further, in the plane of oscillation the bending forces due to the rod's inertia act and weaken it as a strut, while at right angles to the plane of oscillation there are no such forces. Hence the value of m should probably be different for the two cases. If in actual connecting rods of rectangular section, the value of m is the same for bending in both planes, either there is some excess of strength against lateral bending, or the assumption that the connecting rod is fixed at the ends is rather too favourable. Let $h = r b$ where $r = 1.7$ to 2 usually.

For bending in the plane of oscillation $I_1 = \frac{1}{12} h^3 b = \frac{1}{12} r^3 b^4$. For bending at right angles to the plane of oscillation $I_2 = \frac{1}{12} h b^3 = \frac{1}{12} r b^4$. Then, taking a factor of safety of 6 as before, for bending in the plane of oscillation,

$$6 m_1 P = \pi^2 \frac{I_1 E}{l^2}$$

$$b = \sqrt[4]{\frac{72}{\pi^2 E}} \sqrt[4]{m_1} \sqrt[4]{\frac{P l^2}{r^3}} \quad . \quad . \quad . (3b)$$

and for bending at right angles to the plane of oscillation

$$6 m_2 P = 4 \pi^2 \frac{I_2 E}{l^2}$$

$$b = \sqrt[4]{\frac{18}{\pi^2 E}} \sqrt[4]{m_2} \sqrt[4]{\frac{P l^2}{r}} \quad . \quad . \quad . (3c)$$

It appears that in locomotives on the average $m_1 = m_2 = 2$ to 4. For $m_1 = m_2$ the two equations give the same value of b if $r = 2$. Hence let $b = 2h$. We get simply for the rectangular rod

$$b = \sqrt[4]{\frac{9}{\pi^2 E}} \sqrt[4]{m} \sqrt[4]{P l^2} \quad . \quad . \quad . (3d)$$

$m =$	$1\frac{1}{2}$	2	3	4
$\sqrt[4]{\frac{9}{\pi^2 E}} \sqrt[4]{m} =$	0.0147	0.0158	0.0175	0.0188

68. *Diameter of connecting rod for thrust in terms of diameter of cylinder and steam pressure.*—Putting as before $P = \frac{\pi}{4} D^2 p$, and taking for stationary and marine engines

$m = 4$ to 6, we get for wrought-iron or steel rods, of circular section,

$$d = 0\cdot 0341 \text{ to } 0\cdot 0377 \sqrt{\{D l \sqrt{p}\}} \quad . \quad . \quad (4)$$

and for locomotives, with rods of rectangular section, taking $m = 2$ to 4,

$$b = 0\cdot 0149 \text{ to } 0\cdot 0177 \sqrt{\{D l \sqrt{p}\}} \quad . \quad . \quad (4a)$$

$$h = 2b$$

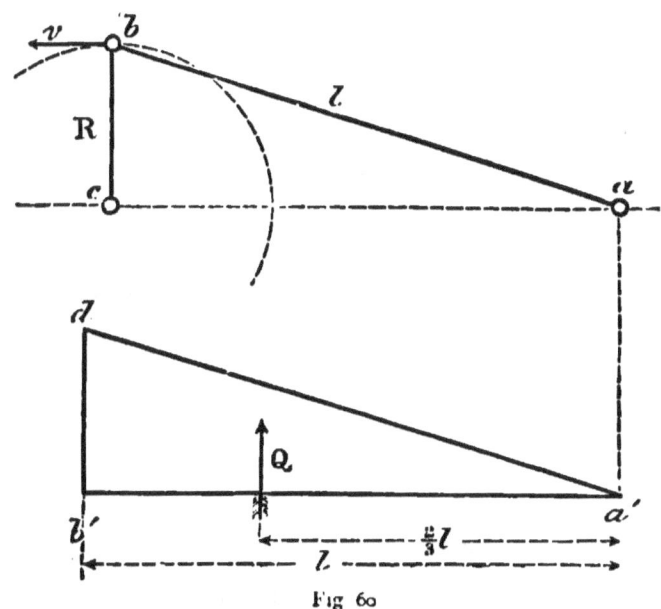

Fig 60

- 69. *Check on the calculation of the strength of the rod, taking into account the forces due to its inertia.*—In the case of fast-running engines it is desirable to check the dimensions obtained by the rule above by calculating the probable stress in the rod due both to the thrust or pull and the bending action due to its own inertia. It is accurate enough to estimate the bending forces due to the oscillation of the rod as if it were a uniform rod of the diameter d at its greatest section, or of the breadth b and height h if the section is rectangular.

Let $\frac{\pi}{4} d^2$ or $b\,h$ be the greatest section of the rod in sq. ins.; l the length between the centres of ends in inches; R the radius of crank in feet, and v the velocity of crank pin in ft. per second. Let 0·28 lbs. be the weight of a cubic inch of iron. Then the assumed weight of the rod is $w = 0\cdot 22 d^2$ lbs. or $0\cdot 28\,b\,h$ lbs. per inch of length.

The tension or pressure f_1 in the rod due to the piston load transmitted is given by the equation

$$\frac{\pi}{4} D^2 p = f_1 \frac{\pi}{4} d^2 \text{ or } f_1\,b\,h \quad . \quad . \quad . \quad . \quad (5)$$

$$\left. \begin{array}{l} f_1 = D^2 p / d^2 \text{ for a circular rod} \\ = \cdot 785\, D^2 p / b\,h \text{ for a rectangular rod} \end{array} \right\} \quad (6)$$

In the position $a\,b$, fig. 60, of the connecting rod, and $b\,c$ of the crank, the acceleration at b is v^2/R. At any point distant x from a the acceleration is xv^2/Rl. The acceleration at the moment is vertical, but it makes no serious error to assume it normal to the rod. The force due to inertia per inch length of rod at x is

$$\frac{w}{g} \cdot \frac{v^2}{R} \cdot \frac{x}{l} = q \frac{x}{l}.$$

Hence we may represent the distribution of load due to inertia along the rod by taking $a^1 b^1 = ab$, setting up $b^1 d = q = \frac{w}{g} \cdot \frac{v^2}{R}$, and joining $a^1 d$. Then the ordinate of the triangle $a^1 d\, b^1$ will be the bending force per inch length of rod. The resultant of the bending forces is $Q = \frac{1}{2} q\,l$ acting at $2/3\,l$ from a^1. The reaction at a^1 is $\frac{1}{3} Q = \frac{1}{6} q\,l$, and that at b_1 is $\frac{2}{3} Q = \frac{1}{3} q\,l$.

At a distance x from a^1 the bending moment is

$$M = q \frac{l x}{6} - q \frac{x}{l} \cdot \frac{x}{2} \cdot \frac{x}{3} = \frac{q}{6 l} (l^2 x - x^3),$$

which is greatest for a value of x given by

$$\frac{dM}{dx} = l^2 - 3x^2 = 0.$$

Hence the bending moment is greatest for $x = l/\sqrt{3} = 0.577\, l$. Its value at that section is

$$M = \frac{1}{16} q\, l^2$$

$$= \frac{1}{16} \frac{w}{g} \cdot \frac{v^2}{R} l^2.$$

The modulus of the section is $z = 0.1 d^3$ or $bh^2/6$. Consequently the stress due to bending is

$$f_2 = \frac{M}{z}$$

$$= 0.137 \frac{v^2 l^2}{g R d} \quad \text{if the rod is circular} \quad . \quad (7)$$

$$= 0.105 \frac{v^2 l^2}{g R h} \quad \text{if the rod is rectangular} \quad . \quad (7a)$$

Hence the total stress in the rod is $f_1 + f_2$, and this should not exceed 5,000 to 6,000 lbs. per sq. in. for a wrought-iron rod, or 6,000 to 7,500 lbs. if the rod is of steel.

For a coupling rod connecting two equal cranks the greatest bending moment is at the centre of the rod, and its value is

$$M = \frac{1}{8} q\, l^2 = \frac{1}{8} \frac{w}{g} \frac{v^2}{R} l^2,$$

and this is to be substituted for the value of M in the equations above.

One end of a connecting rod is sometimes forked, so as to carry two journals. Then the forging is more complicated, and the brasses and cotters are doubled in number. With a forked rod it is almost impossible to adjust the brasses so that there is equal thrust at each of the journals. There is then a lateral thrust on the crosshead, and the connecting rod is strained by the bending due to the thrust deviating from its axis. The brasses wear unequally also. The forked rod is therefore mechanically bad, and if it must be adopted it should at least have extra strength.

CONNECTING-ROD ENDS

70. Connecting-rod ends for steam engines are commonly termed 'big ends' and 'little ends,' the former being the bearing for the crank pin and the latter for the crosshead pin. The crosshead pin being a neck journal, usually, and subject only to wear due to rotation through a small angle, is generally a good deal smaller than the crank pin. In marine engines with cranked shafts, the crank-pin journal is very large compared with the crosshead pin.

71. *Proportions of steps.*—The ends of connecting rods are designed to receive crank pins or neck journals, and are fitted with gun-metal steps similar to those used for pedestals. (Part I., Chap. VIII.) The unit for the proportional numbers relating to the steps in connecting rods is

$$t = 0.08\,d + \tfrac{1}{8} \quad . \quad . \quad . \quad . \quad (8)$$

where d is the diameter of an ordinary crank pin supporting the thrust transmitted by the connecting rod. When the connecting rod is attached to a journal of greater size than is sufficient for the thrust P_1 of that connecting rod only,

$$t = .007\sqrt{P_1} + \tfrac{1}{4} \text{ to } .012\sqrt{P_1} + \tfrac{1}{4} \quad . \quad (9)$$

The flanges of the steps are of very variable thickness, but very often the space between the flanges of the steps in which the connecting-rod end is placed is $\tfrac{7}{10}$ths of the length of the journal.

72. *Strength of connecting-rod ends.*—The piston load being P, calculated as above, the greatest straining force acting axially along the rod is $m\,P$, where, as has been seen, m is about $1\tfrac{1}{4}$ to $1\tfrac{1}{2}$ in different cases.

The strap is subject to tension only. It may therefore carry a working stress f of about 10,000 lbs. per sq. in. for wrought iron or 13,200 lbs. for steel. Its breadth being β and thickness $\bar{c}$, fig. 61,

$$\beta\bar{c} = \frac{m\,P}{2f} \qquad . \qquad . \qquad . \qquad . \qquad . \qquad . \quad (10)$$

$\beta\bar{c}$ = 0·0000625 to 0·0000750 P for wrought iron
  = 0·0000473 to 0·0000568 P for steel.

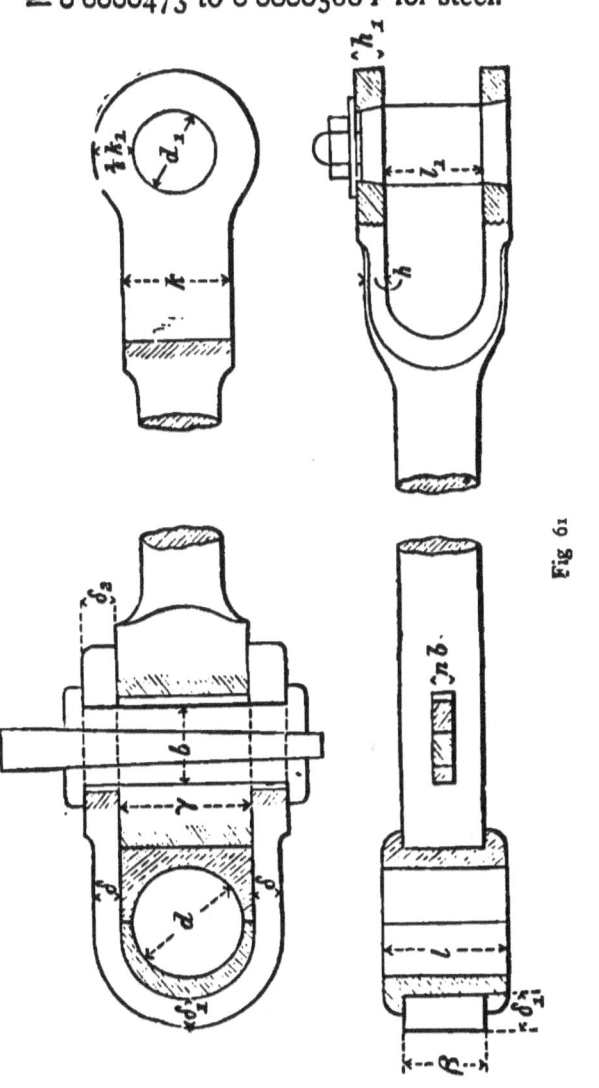

Fig 61

The thickness c_1 is taken 1·2 to 1·5 δ to allow for the undeterminable bending action which arises out of the strap tightening down on the brass when the load comes on. The thickness c_2 is taken about 1·3 δ partly to make up for the metal cut out to form the cotter hole and partly to give bearing surface enough for the crushing pressure of the cotter.

The gibs and cotter should be proportioned by the rules in Part I., p. 179. If b is the width of gibs and cotter, and nb their thickness, then usually $n = \frac{1}{4}$. Here the stress on the cotter acts in one direction only. For the shearing section

$$2 n b^2 f_s = m P$$

$$b = \sqrt{\frac{m}{2 n f_s}} \sqrt{P} \qquad . \qquad . \quad (11)$$

Putting $n = \frac{1}{4}$, $m = 1\frac{1}{4}$ to $1\frac{1}{2}$, $f_s = 8{,}000$ for wrought iron and 13,200 for steel.

$$\sqrt{\frac{m}{2 n f_s}} = 0\cdot00442 \text{ to } 0\cdot00484 \text{ for wrought iron}$$
$$= 0\cdot00344 \text{ to } 0\cdot00377 \text{ for steel.}$$

For the other end of the rod the total section through the fork is $2 k h$, and as this is subjected alternately to tension and pressure we should take $f = 5{,}000$ for wrought iron and 6,600 for steel. Then

$$k h = \frac{m P}{2 f} \qquad . \qquad . \qquad . \qquad . \qquad . \quad (12)$$
$$= \cdot000125 \text{ to } \cdot000150 \text{ P for wrought iron}$$
$$= \cdot000095 \text{ to } \cdot000114 \text{ P for steel.}$$

The section $k_1 h_1$ through the pin is increased to about $1\frac{1}{4} k h$ to allow for bending stresses due to the distribution of the load over the inner surface of the eye.

73. *Forms of connecting-rod ends.*—Fig. 62 shows a very common form of connecting-rod big end, having a loose strap confining the brasses, kept in place by gibs and cotter.

It will be seen that tightening the cotter shortens the connecting rod. The strap-end form is not well adapted for high-speed engines, because the transverse forces due to the inertia of the rod tend to open the strap. The strap is of wrought iron or steel. Its section may be determined by

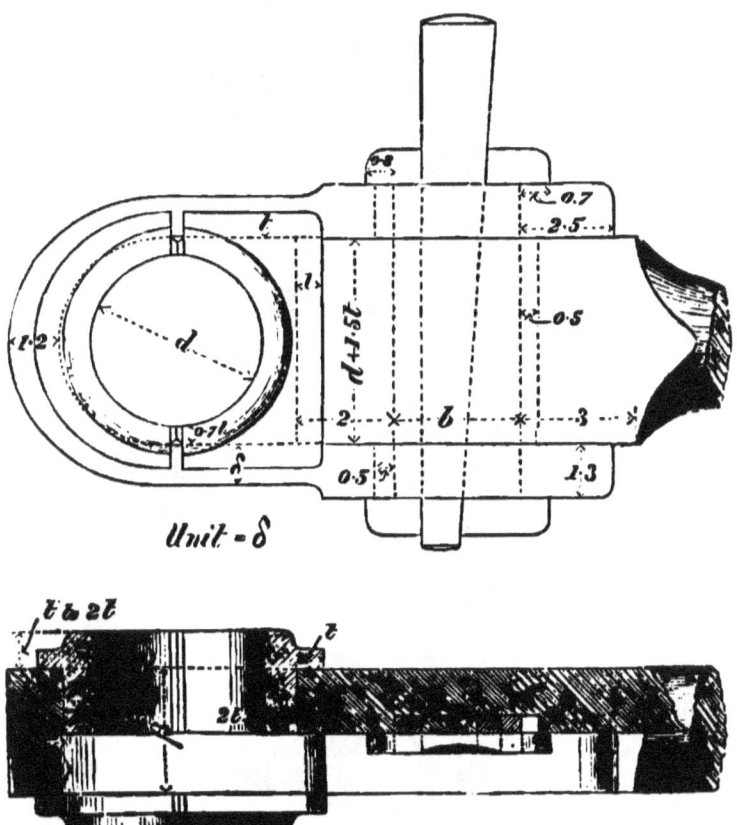

Fig. 62

the rule above, but cases will be found in practice where fifty or a hundred per cent. more section has been given.

In the figure a proportional unit $\delta = 0\cdot15\,d + 0\cdot2$ has been taken. The inclination of the sides of the cotter may be $\frac{1}{30}$ to $\frac{1}{15}$ on each side, if there is no locking arrangement, and $\frac{1}{6}$ if there is a set screw.

Although it is very convenient to design machine parts such as these by proportional figures, it is always desirable

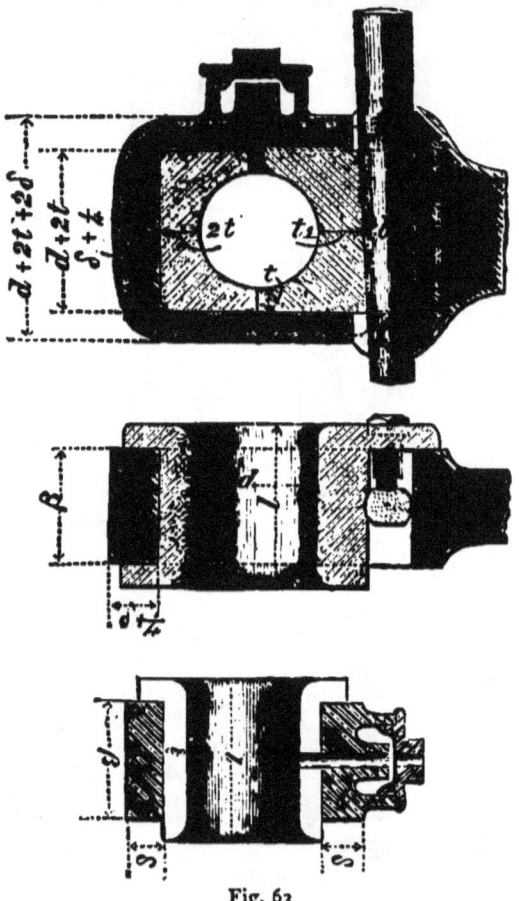

Fig. 63

to check the dimensions of the most important sections by calculating the stress on them. The stress should not exceed the values in Table II., Part I., after ample margin has been left for straining actions neglected. Saving of weight is not very important in such machine parts as these, and very often in practice a very large factor of safety will be found, especially in small engines.

74. *Box end.*—Fig. 63 shows a connecting-rod end

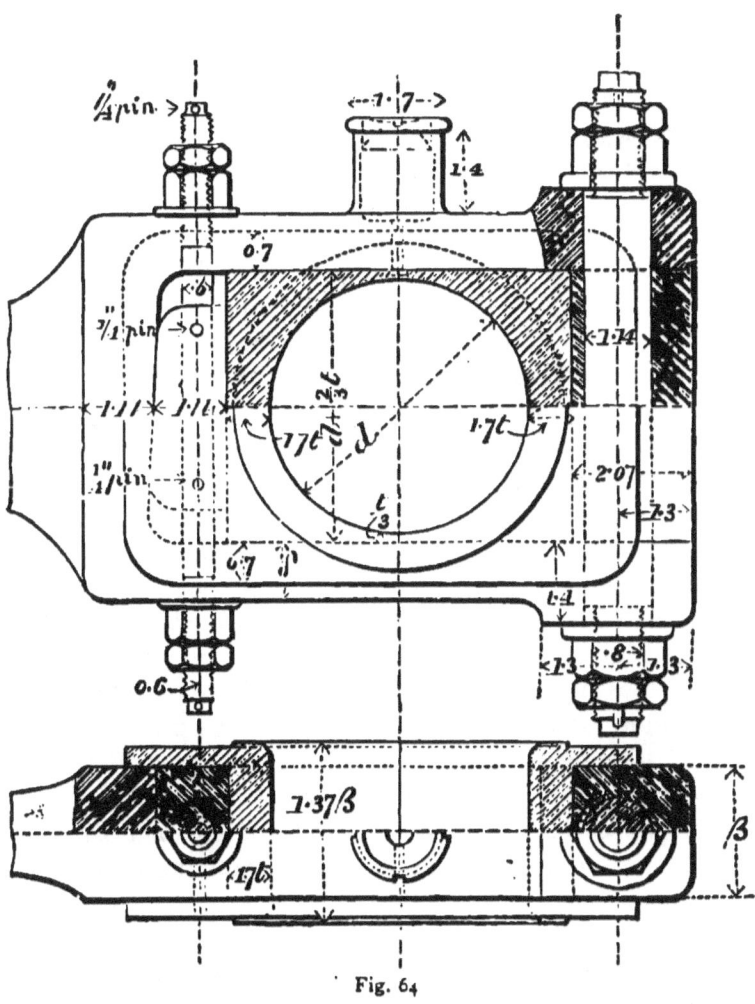

Fig. 64

having no loose strap. The brass steps have a thickness $2\,t$ opposite the key, and $t_1 = 6\,t - \tfrac{1}{2}$ next the key. At the

II.

sides the thickness is reduced to t. The thickness and overlap of the flanges of the steps may be $\frac{1}{6}l - \frac{1}{8}$, so that the width of the box may be $\beta = \frac{3}{3}l + \frac{1}{4}$. The flanges of the steps are partially removed on one side to allow their insertion in place. The thickness $\bar{c}$ of the sides of the box may have the same value as the thickness of strap in the last case. The mean breadth of the cotter is $0.6\,\beta$ and its thickness $0.3\,\beta$, and it tapers 1 in 12 on each side. It is secured by two set screws; diameter of set screws = cotter thickness. Unlike the last form, this connecting rod is lengthened when the cotter is tightened. But it may be arranged with the cotter on the other side of the brass steps, and then it is shortened by tightening the cotter. A coupling rod should have one end arranged in the former and one in the latter method. Then the length of the rod is not much altered by tightening the cotters. The proportional unit for this figure is $\bar{c} = 0.15\,d + 0.2$.

Fig. 64 shows a locomotive connecting-rod end which, whilst it is adjusted like a box end, can be separated from the shaft like a strap-ended rod. The block forming the end is held in place by a stout pin having a slight taper, with nuts at each end. The wedge is tapered 1 in 16, and is fixed by two peg pins driven through the adjusting bolt. The ends of the adjusting bolt are left long, and have double nuts at each end to facilitate the adjustment of the position of the wedge. The flanges of the steps are large, to cover the spaces in the wrought-iron frame of the end. In this case $\beta = 2.3\bar{c}$, for the proportions given on this rod, t is obtained by equation 8, and the unit for the other proportional numbers is $\bar{c}$. The value of m is about 3.12.

75. *Marine engine connecting-rod end.*—Fig. 65 shows another form of connecting-rod end. This is of simple and massive form, and is often used in marine engines. The brasses are lined with white metal or Babbitt's metal cast in shallow recesses.

Connecting Rods

If δ_1 is the diameter of the bolt at the bottom of the screw thread and f the working stress,

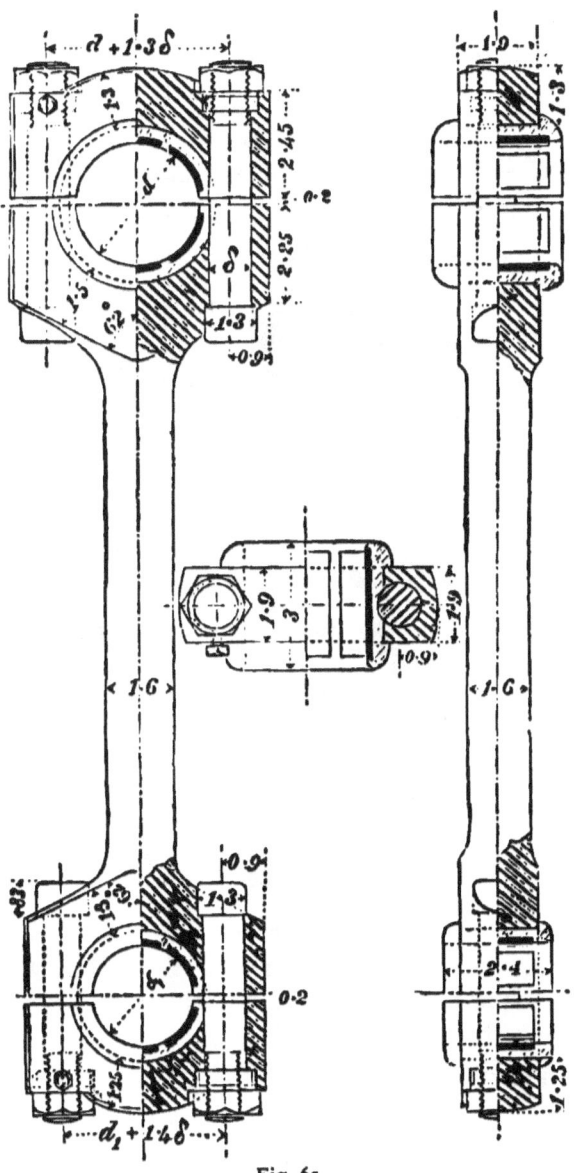

Fig. 65

$$\frac{\pi}{4} \delta_1{}^2 f = m \, P.$$

The diameter δ at top of threads will be nearly enough for the present purpose $1 \cdot 12\, \delta_1$. Hence

$$\delta = \sqrt{\frac{1 \cdot 25\, m}{\cdot 785\, f}} \sqrt{P} \qquad . \qquad (13)$$

Putting $f = 5,000$ for wrought iron and 6,600 for steel

$$\delta = 0 \cdot 0200 \text{ to } 0 \cdot 0218 \sqrt{P} \text{ for wrought iron}$$
$$= 0 \cdot 0174 \text{ to } 0 \cdot 0190 \sqrt{P} \text{ for steel.}$$

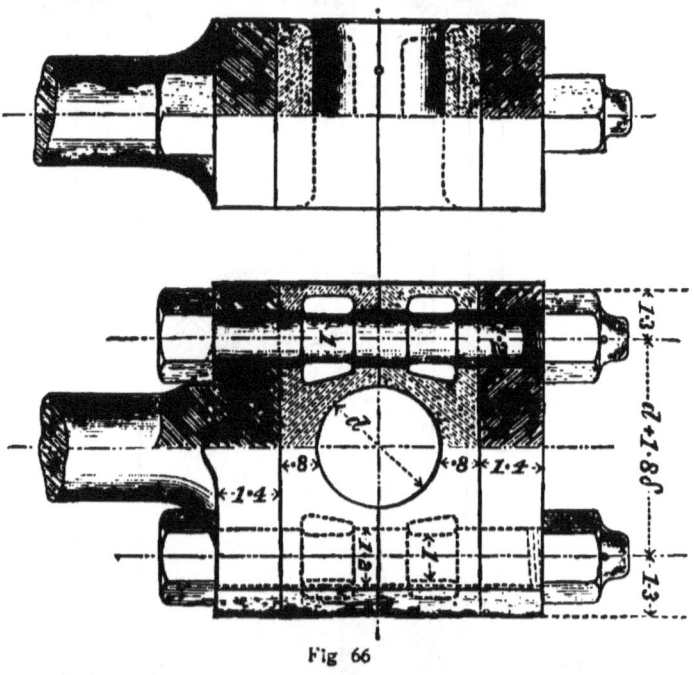

Fig 66

It is a good thing to turn part of the shank of the bolt to a diameter equal to that of the bottom of the screw thread. This gives the bolt a little elasticity without weakening it. The proportional numbers in the figure are reckoned to the unit $7/8\, \delta$ or δ_1.

Fig. 66 shows a somewhat similar big end for a connect-

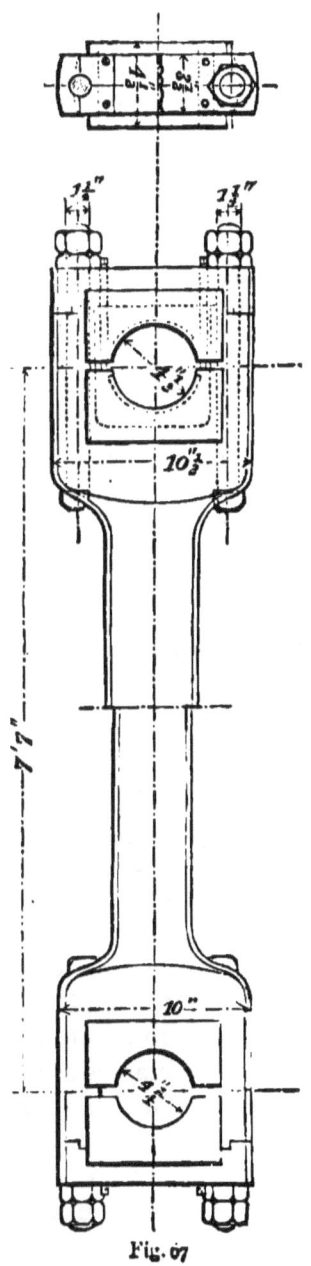

Fig. 67

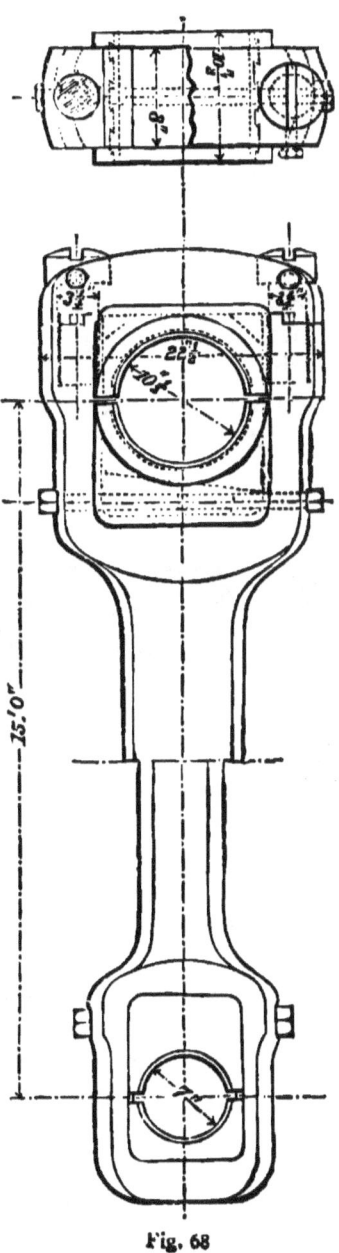

Fig. 68

ing rod. Here the brasses are designed so that they can be machined all over and are cored internally to diminish weight. The bolts are turned down so as to bear only at those points where they most effectively support the brasses and caps and prevent lateral movement.

Figs. 67, 68 show connecting rods designed by Mr. W. F. Mattes.[1]

In fig. 67 the end has jaws long enough to hold both brasses. The cap hooks over projections on the jaws, turned concentrically with the rod, and the cap is bored to fit the projections. Thus the bolts are relieved of any straining action except the longitudinal tension, which they are best capable of resisting. The cap is lighter than in the

Fig. 69

ordinary marine-engine type. As compared with a cap secured by a transverse bolt, as in fig. 64, the forging of the head is simpler and the bolts in tension are more secure than the eyes in the other form which hold the bolt in shear. As compared with a solid-head rod, the open end is frequently more convenient. In fig. 67 the wear is intended to be taken up by liners. In fig. 68 a wedge is introduced, and tapped bolts are used instead of through bolts to secure the cap. Bolts and rod are of forged steel. The brasses are of forged steel. The brasses are of cast steel faced with Babbitt metal.

76. Fig. 69 shows a connecting-rod small end, from a design by Mr. Halpin. Part of the pin on each side is planed away

[1] 'Trans. Am. Soc. Mechanical Eng. ix. p. 467.

and the brasses undercut. As the pin only vibrates through a small angle, it overruns the edges of the brasses, which prevents shoulders being formed by wear. The spaces allow lubricant to reach the wearing surfaces easily and the difficulty of an oval pin is avoided. Solid bushes instead of two steps are sometimes used for connecting-rod small ends. They are cheaper, and the wear is not very great if there is good lubrication.

Fig. 70 shows a coupling-rod joint which may serve as an example of a journal bearing where there is not a great

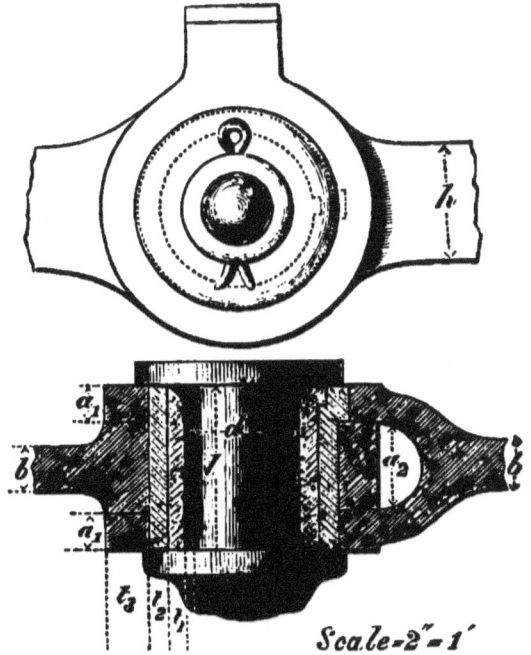

Fig. 70

amount of motion and wear. This joint is intermediate in construction between a common knuckle-joint and a connecting-rod end. It has bushes to diminish friction and wear, but these are not divided, so that there is no adjustment after wear has taken place. The crank pin turns in a brass bush, which is protected by an outer steel bush. Both brass and steel bush are fixed in the forked-rod end by small snugs, and the solid-rod end turns on the steel bush. The pin in a joint of this kind is often larger than is necessary for strength, because, by using a large pin with a small intensity of pressure between the rubbing surfaces, there is less danger of squeezing out the lubricant. The pin is of steel. The proportions may be

$$t_1 = t_2 = 0{\cdot}1 d + \tfrac{1}{8}$$
$$a_1 = 0{\cdot}3\, d.$$
$$a_2 = 0{\cdot}8\, d$$
$$t_3 = \frac{b\, h}{4\, a_1}.$$

For the brasses of parallel motion bars, which should be capable of being tightened without altering the position of the centres of rotation, the following ingenious plan has been suggested by Mr. Candlish ('Engineering,' xxxii. 461). The motion bars are fitted with bushes tapered internally to fit conical journals, and parallel externally, with a feather to prevent the bushes from turning in the rod eyes. By regulating the position of the bushes by double nuts they can be adjusted when worn without interfering with the centres of motion.

CHAPTER V

CROSSHEADS AND SLIDES

Crossheads

77. 'Crosshead' is the name given to the part which connects together the piston rod and connecting rod of a steam engine, and with which is also connected the guiding arrangement either of slide blocks or parallel motion bars. It consists essentially of a socket to which the piston rod is keyed, and a crosshead pin, forming one or more journals on which the connecting rod works. The crosshead either forms a slide block or has the slide blocks attached to it.

The design of the crosshead depends primarily on the arrangement of the slides which guide the piston-rod end. In the older engines and some modern engines there are two slides (formed by four slide bars), one on each side of the crosshead. Now frequently there is only one slide formed by slide bars above and below the crosshead, or the crosshead is formed into a slipper slide block guided in a channel on one side of the crosshead.

The crosshead pin is most commonly fixed in the crosshead. If the latter is forked, the pin forms a neck journal, on which the connecting rod works, fig. 72. In other cases the connecting rod is forked and works on journals on each side of the crosshead, fig. 73. A third arrangement is to fix the crosshead pin in a fork of the connecting rod. Then brasses must be arranged in the crosshead, fig. 77.

The sliding surfaces which receive the lateral thrust of the connecting rod and prevent bending of the piston rod are sometimes formed on the crosshead itself, fig. 75; sometimes on slide blocks attached to the crosshead by journals, fig. 72. If the wearing surfaces of the slide blocks are large there is sometimes no adjustment for wear. Sometimes a simple adjustment is obtained by putting thin washers or lining pieces in the supports of the slide bars; by removing these the slide bars are made to come closer together. More generally, in modern engines the slide block has one or more adjusting wedges fixed by screws or cotters.

78. *Forces acting at the crosshead.*—Let D be the diameter of the cylinder, p the effective steam pressure, then the pressure on the crosshead due to the piston load is

$$P = \frac{\pi}{4} D^2 p \qquad . \qquad . \qquad . \qquad (1)$$

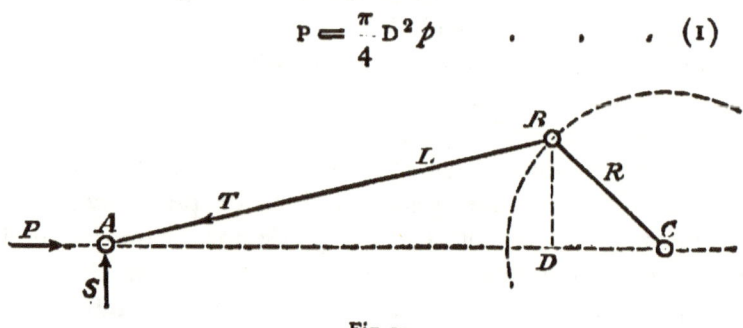

Fig. 71

This is balanced at the crosshead by a thrust T along the connecting rod and a lateral reaction S due to the pressure on the slide blocks. Let R be the crank radius; $L = nR$ the connecting rod length; θ the crank angle with the line of stroke. Then

$$P : T : S :: AD : AB : BD$$
$$:: \sqrt{n^2 - \sin^2 \theta} : n : \sin \theta$$

If we treat P as constant through the stroke

$$S = P \frac{\sin \theta}{\sqrt{(n^2 - \sin^2 \theta)}} \qquad . \qquad . \qquad . \qquad (2)$$

This is a maximum when ABC is a right angle, and then the greatest pressure on the slide block is

$$S_{max} = P/n \qquad . \qquad . \qquad . \qquad . \quad (3)$$

Lengthening the connecting rod diminishes this pressure, and that is one of the reasons for adopting a long connecting rod. If values of s are set up along the stroke they form an approximate ellipse. Hence with a constant piston load, the mean pressure on the slide bars is

$$S_{mean} = \frac{\pi P}{4 n} \qquad . \qquad . \qquad . \quad (4)$$

Usually, however, the steam pressure diminishes through the stroke and S_{max} and S_{mean} have values less than those given above. There are also the inertia forces which have been neglected. These do not much affect the greatest slide-block pressure, which occurs near midstroke.

79. *Crosshead pin.*—Most commonly the crosshead pin forms a neck journal to receive the thrust of the connecting rod. This thrust is affected considerably both by the variation of steam pressure and the inertia forces As in the case of connecting rods, the greatest thrust may be taken to be

$$T = m P,$$

where m is a factor ranging from $1\frac{1}{4}$ to $1\frac{1}{2}$. The velocity of rubbing at the crosshead pin is small, and the pin is often of steel or case-hardened wrought iron. Hence a high intensity of pressure is permitted per sq. in. of bearing surface of the journal. By the rules in Part I., Case II. *a*, p. 201, if d is the diameter and l the length of the journal, p the intensity of pressure on bearing surface and f the safe stress,

$$\left. \begin{array}{l} d = \sqrt[4]{\dfrac{1 \cdot 28}{pf}} \sqrt{m P} \\ l = \dfrac{m P}{p d} \end{array} \right\} \qquad . \quad (5)$$

It appears that p is often 1,200 to 1,500 lbs. per sq. in. and f

is generally not more than 5,000 to 6,000 lbs. per sq. in. for wrought iron or 6,600 to 7,500 for steel. Then

$$d = 0\cdot0215 \text{ to } 0\cdot0194 \sqrt{m\ P} \text{ for wrought iron}$$
$$= 0\cdot0200 \text{ to } 0\cdot0184 \sqrt{m\ P} \text{ for steel.}$$

It is common in many cases to make the length equal to the diameter of a crosshead pin. Then

$$\left. \begin{array}{l} d = \frac{4}{7} \sqrt[4]{\dfrac{1\cdot28}{pf}} \sqrt{m\ P} \\ = \frac{4}{7} \sqrt{\dfrac{m\ P}{p}} \end{array} \right\} \quad . \ (6)$$

80. *Forms of crosshead.* —Fig. 72 shows a simple crosshead for an arrangement of four slide bars. The crosshead

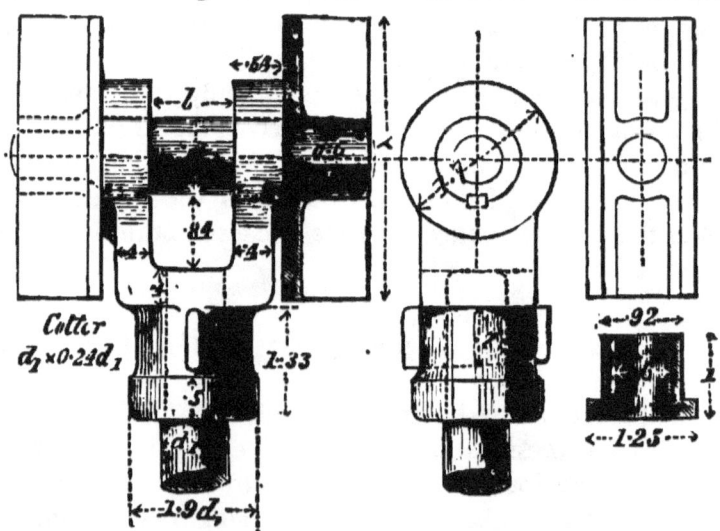

Fig. 72

is of wrought iron, cottered to the piston rod, and having a forked end embracing the connecting rod. A pin passing through the crosshead forms a neck journal for the connecting rod, and at the same time two end journals on which the slide blocks are fixed. The slide blocks are simple cast-iron blocks. In large engines these blocks have brass faces on the rubbing surfaces. The pin must be fixed in the jaws of the crosshead by a small key, shown in the end view, which prevents the rotation of the pin. For the connecting-rod journal $d = l$. The unit for the proportions of the other parts of the crosshead is d.

Fig. 73 shows the form of the crosshead pin when the crosshead has a single end and the connecting rod is forked. Each connecting-rod end is designed as above described, but for half the total thrust in the rod.

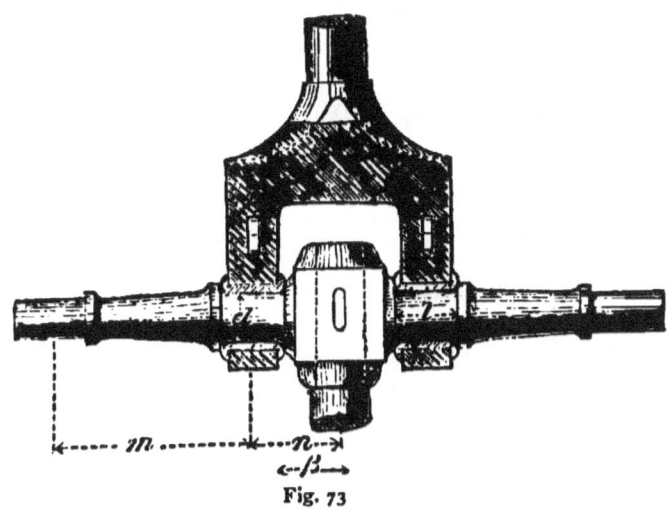

Fig. 73

Fig. 74 shows a simple crosshead equivalent to that in fig. 72, but arranged with two slide bars only, above and below the crosshead. The slide blocks are of cast iron, and the slide bars of steel. The design has, however, a fault not uncommon in crossheads. In order that the vertical pressure may be uniformly distributed over the slide block, the centre of the crosshead pin should be over the centre of the slide block. If it is not so, either the pressure is very unequal, or it is only prevented from being so by the stiffness of the piston rod. Besides the bad distribution of the pressure causing increased wear, it tends to force out the lubricant.

Figs. 75, 76 show two forms of crosshead applicable when there are two slide bars in the plane of oscillation of the connecting rod. The piston-rod socket is proportioned to the piston-rod diameter, d_1. In both these ex-

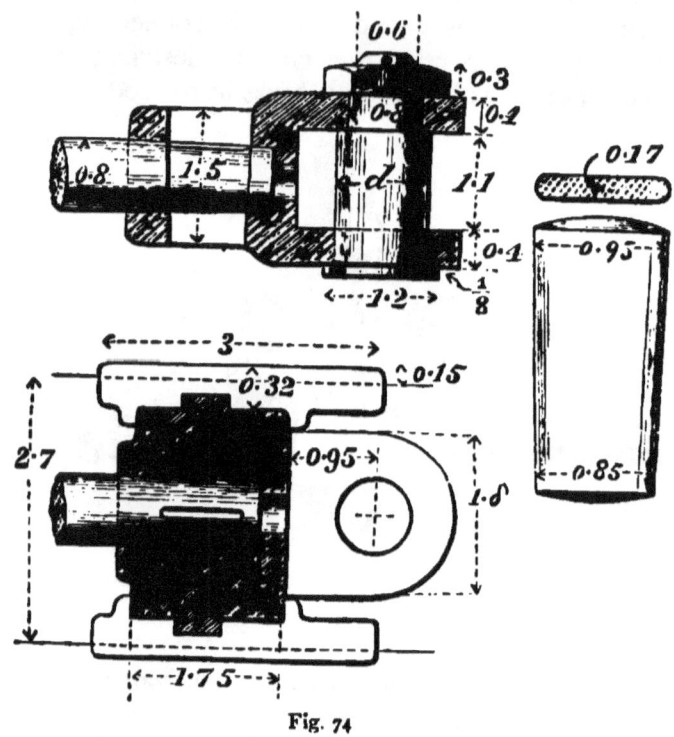

Fig. 74

amples the piston rod is enlarged at the crosshead end. This involves a split stuffing-box. The unit for the remaining parts is the crosshead pin diameter, d. In fig. 75 the crosshead is entirely of wrought iron, except the brass faces attached by set screws to the rubbing surfaces. The crosshead pin is kept in place by a **T**-headed bolt, which passes completely through it. The ends of the pin are tapered, and rotation of the pin is prevented by friction of the tapered parts.

In fig. 76 the crosshead of wrought iron and the slide

Fig. 75

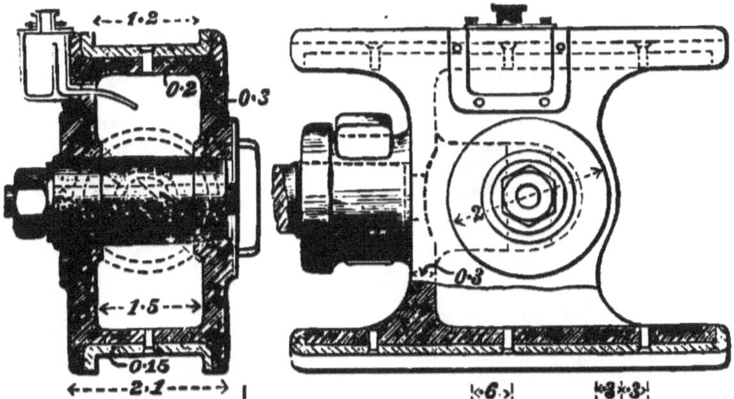

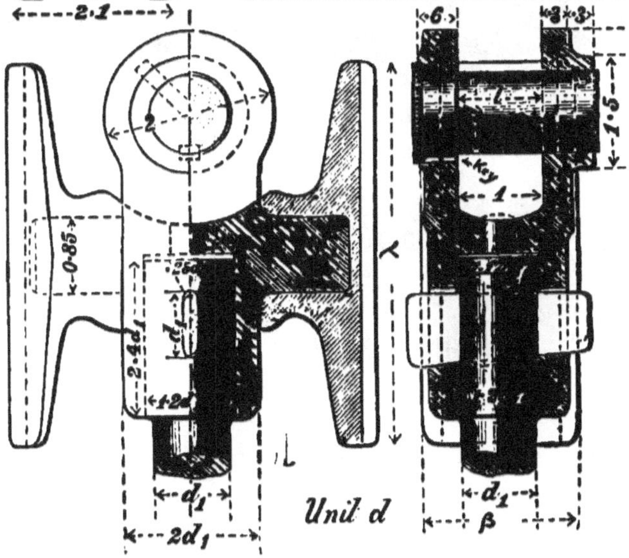

Fig. 76

blocks are separate, and of cast iron. The crosshead pin is kept in place by a split pin, and rotation is prevented by a small key inserted on one side.

Fig. 77 shows a crosshead designed by Mr. Stroudley for a slipper slide. The crosshead is of wrought iron forged in one piece with the piston rod and slide block, and the connecting rod is forked at the end, and embraces the crosshead. The steps of the crosshead pin are of gun-metal, or of case-hardened wrought iron, and are tightened by a wedge and set screw. The crosshead pin, of case-

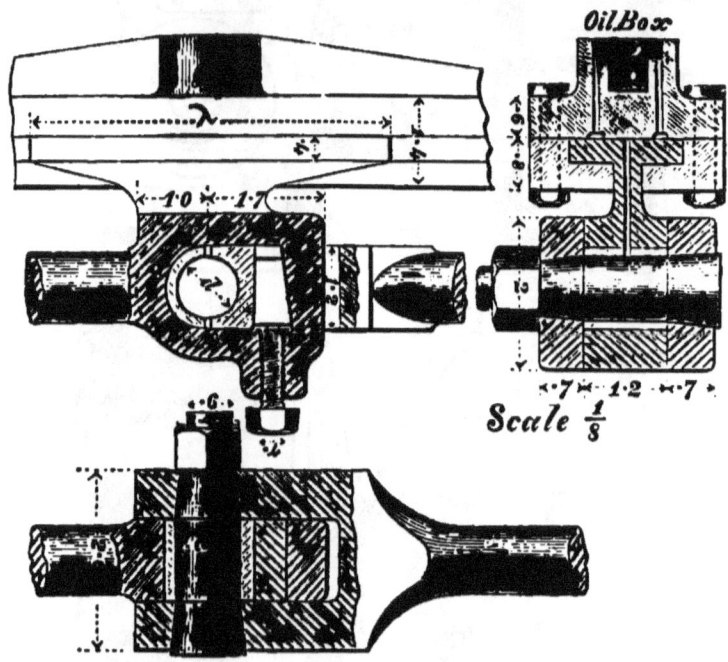

Fig 77

hardened wrought iron, is fixed to the jaws of the connecting rod. In this case the slipper slide block is over the crosshead. The reason of this is that in locomotives the principal pressure on the slide, when the engine is running forwards, is upwards.

Figs. 78, 78*a* show a crosshead for a single slide bar, designed by Mr. Adams for the engines of the Great Eastern Railway. The slide block is in two parts, of cast iron. A

Crossheads and Slides

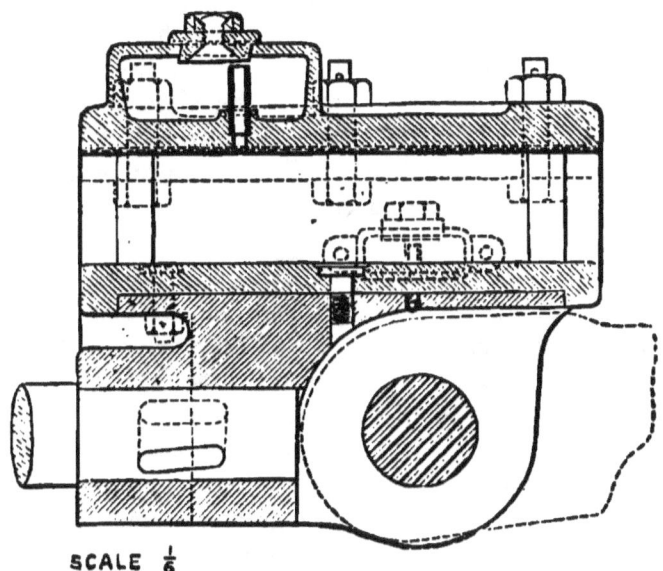

SCALE 1/6

Fig. 78

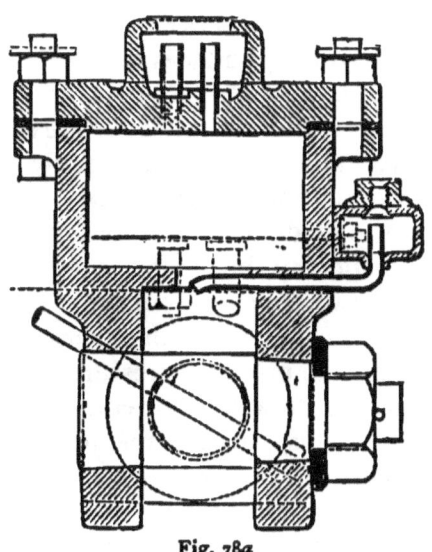

Fig. 78a

projection below receives the end of the piston rod cottered in a conical hole. Forked jaws carry the crosshead pin. The connecting-rod small end is single with a solid bush. Six $\frac{7}{8}''$ bolts connect the two parts of the slide block. The slide bar is 8" × 3" of steel with holes through it to permit the oil to reach the under side. The pressure on the slide block is about 40 lbs. per sq. in.

SLIDE BARS

81. In most linkwork arrangements it is necessary to guide the ends of some of the bars, so as to constrain them to move in straight lines. This can be done by an arrangement of links forming what is termed a 'parallel motion.' Into the construction of parallel motions no elements enter which have not already been discussed. A parallel motion may be made to guide a given point with great accuracy and with very little friction. On the other hand, it is from the point of view of mechanical construction a somewhat complicated arrangement, and if the links alter in length by wear it no longer properly answers its purpose. Hence, parallel motions have been to a great extent superseded by a simpler arrangement of straight-guiding surfaces termed 'slides.' Slide bars and slide blocks waste more work in friction than parallel motions, unless, indeed, the latter are out of adjustment. But they are more cheaply and easily constructed, and the waste of work is not serious. Hence, slides are now almost always used to guide the end of piston rods.

In ordinary stationary horizontal engines the crank throws over in the forward stroke from the cylinder towards the crank shaft and passes under the crank shaft in the return stroke. Then the principal pressure on the slides is downwards both in the forward and return stroke. There is some advantage in this, because the lower slide bar is most easily lubricated. In locomotives, since the cylinders are forward of the driving axle, the reverse condition holds. The principal pressure is upwards in forward running, and is only downwards when the engine is reversed. In marine engines the slide-bar surfaces, which take the principal pressure when going ahead, are often made larger than those which have to take the pressure when going astern.

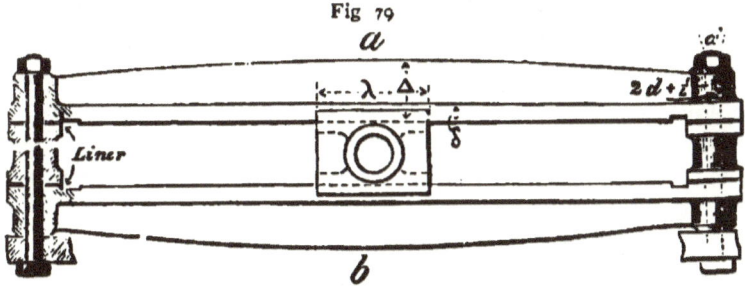

Fig 79

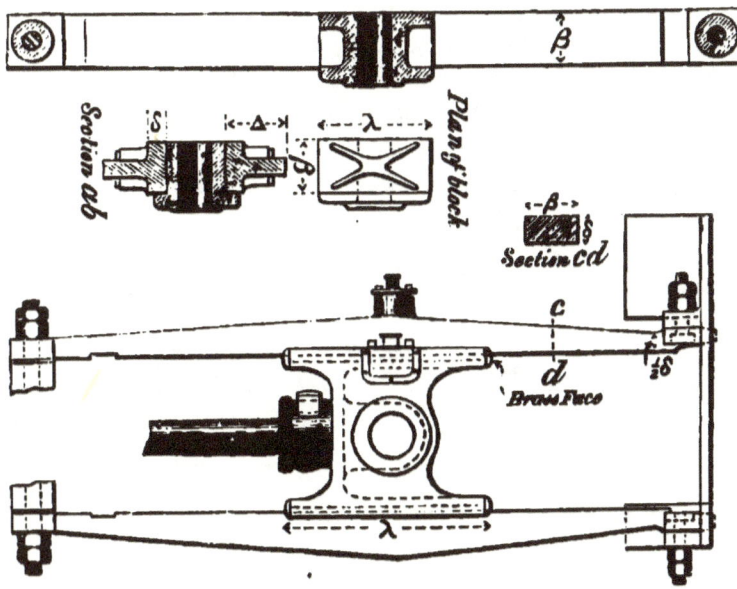

Fig 80

Fig. 79 shows a pair of ordinary slide bars with the slide block between them. The bars are of T section and are spaced apart at the ends by distance pieces. Thin washers of liners are introduced between the distance pieces and the bars, so that, when the bars and slide block are worn, the bars can be brought closer together.

The bars are notched at the ends, and the slide block passes the edge of the notch at each stroke. This prevents the formation of a ridge at the end of the stroke, in consequence of the wear of the bar. Ample provision must be made for lubricating the bars. The slide blocks may be of cast iron or of gun-metal. They fit on journals at the end of the crosshead pin. The arrangement will be understood, if the crosshead and side blocks in fig. 72 are compared with the slide block and slide bar in fig. 79.

Fig. 80 shows slide bars of wrought iron (sometimes case-hardened) or steel slide bars. In the arrangement here shown the bars are above and below the crosshead. The bars are rectangular in section, and thickest at the centre where the thrust is greatest. The crosshead here shown has brass faces. With this kind of arrangement, in large engines, provision is made to neutralise the wear of the bars by separating the surfaces of the slide blocks.

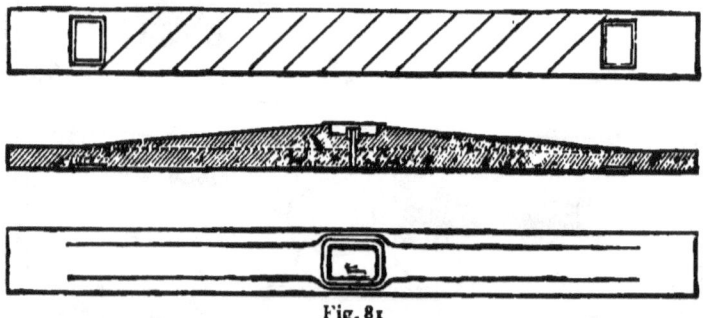

Fig. 81

Fig. 81 shows an American slide bar of cast iron [1] partly chilled on the surface. The chilling has been effected in diagonal strips, leaving equally wide spaces of unchilled cast

---

[1] Rigg, 'Steam Engine,' p. 285.

iron between. The arrangement seems well adapted to secure good lubrication, and at the same time a sufficient area of very hard surface.

A slipper slide is sufficiently shown in the arrangement already described (fig. 77). The slide block is a T-shaped piece, forged in one with the crosshead, and this is guided in a groove formed by a flat slide bar and two L-shaped bars. Other forms of slide bar are used, the sliding surfaces being sometimes wedge-shaped and sometimes cylindrical.

82. *Wearing surfaces of slides and slide blocks.*—Slides and slide blocks should be so designed that the wear is very small, because adjustment is always troublesome. It has been shown above that the mean pressure on the slide block, normal to the sliding surface, is $\pi P/4 n$, under the assumption that the variation of steam pressure and the inertia forces are disregarded. Now let μ be the coefficient of friction and v the mean velocity of the piston and crosshead. Then the work wasted in friction is $U = \mu \dfrac{\pi}{4} \dfrac{P}{n} v$ foot lbs. per second. Both the wear and the heating must be supposed proportional to U. Now let a be the area of the slide block (or two slide blocks) supporting the lateral thrust of the connecting rod and let t be the thickness worn off in any given length of time. Then

$$a\,t \text{ varies as } \mu \dfrac{P}{n} v,$$

and for any given depth of wear a should be proportional to the piston load P and piston speed v, and inversely as the ratio of connecting rod to crank. We should arrive at the same result if we consider that h thermal units were conducted or radiated from each unit of slide-block surface per second and that the rise of temperature should be limited.

It has been usual to design the slide-block surface with reference to the maximum pressure P/n calculated from the greatest piston load due to the initial steam pressure. But even so great discrepancies occur in different cases in practice. Mr. Rigg has given a series of cases ('Treatise on the Steam Engine,' p. 124) of slide blocks of stationary, marine, and locomotive engines in which the slide-block pressure ranges from 22 to 126 lbs. per sq. in. In marine engines the surface for going astern, which is comparatively seldom used, is so small in some cases that the pressure is 400 lbs. per sq. inch.

If β is the width and λ the length of the slide block, then the area supporting the pressure is $a = \beta\lambda$, if there is one slide block and $a = 2\beta\lambda$ if there are two slide blocks. This area a should be so arranged that

$$a = \frac{P}{ny} = \frac{\pi}{4}\frac{D^2 p}{ny},$$

where in good practice y ranges from 30 to 50 in stationary engines, from 40 to 60 in good locomotives, and from 40 to 100 in large marine engines. The lower values should be chosen when the speeds are high.

The slide-bar surfaces are usually of cast iron, except in locomotives, where steel is often used. The rubbing surfaces of the slide blocks are generally of cast iron, sometimes of gun-metal. Oil-grooves are formed on the slide-block surface, and sometimes shallow grooves are planed across it to hold lubricant. These depressions are not reckoned as diminishing the slide-block surface.

83. *Strength of the slide bar.*—Let m and n be the distances from the centre of the connecting-rod eye to the points of support of the slide bar, when the crank and connecting rod are at right angles. Then the greatest bending moment on the slide bar, immediately under the connecting rod, is

$$P\frac{r}{l} \cdot \frac{m\,n}{m+n} \qquad . \qquad . \qquad . \qquad (7)$$

Hence, if the section of the bar is rectangular of breadth β and thickness δ,

$$\tfrac{1}{6}\beta\delta^2 f = P\frac{r}{l} \cdot \frac{m\,n}{m+n},$$

$$\delta = \sqrt{\left\{ \frac{6}{f} \cdot \frac{r}{l} \cdot P \frac{mn}{(m+n)\beta} \right\}}$$
$$= k \sqrt{\left\{ P \frac{mn}{(m+n)\beta} \right\}} \quad . \quad . \quad (8)$$

The limiting stress should be taken at 6,000 lbs. for wrought iron or steel, to allow for the straining actions due to reaction and to secure stiffness; and at about 3,000 lbs. for cast iron. Hence,

$\frac{l}{r} =$	$3\frac{1}{2}$	4	5	6	
$k =$	·0169	·0158	·0141	·0129	for wrought iron,
=	·0239	·0224	·0200	·0183	for cast iron.

The T-shaped section for cast iron is more rigid, but not much stronger than if the feather were omitted.

When the slide bars are horizontal, the weight of the connecting rod, crosshead, &c., rests on the lower bar. If, then, the engine runs only in one direction, it may be arranged so that the thrust due to the pressure transmitted acts on the upper bar, provided at least that the crank is driven by the piston, and that the crank does not for part of the stroke drag the piston. Then the weight and thrust partially neutralise each other, and friction and wear is diminished. If the engine runs in both directions, but more constantly forwards than backwards, the surface of the slide block, which receives the thrust when running forwards, is often greater than that which receives the thrust when running backwards. This is the case in fig. 77. When the engine runs forward, the thrust is upward; when running backward, the thrust is downward.

The surfaces of slides are usually plane. Sometimes the slide bars are cast on the cylinder, and then it is most convenient to bore out the space for the slide block concentric with the cylinder. Necessarily then the surface of the slide bars is cylindrical.

In slide bars having large surfaces and well lubricated the wear is small. Hence sometimes adjustments to neutralise wear are omitted.

CHAPTER VI

PISTONS AND PISTON RODS

84. A piston, or plunger, is a sliding piece which is either driven by fluid pressure or acts against fluid pressure as a resistance. Pistons and plungers are commonly circular in section, and are guided by cylindrical bearing surfaces, so as to reciprocate in a straight path. But other forms of piston are occasionally used.

A plunger is a single-acting piston—that is, a piston receiving the action of the fluid on one face only—and it is guided, not by the cylinder itself, but by a stuffing-box in the cylinder cover. The bearing surface of the plunger therefore requires to be longer than the stroke. The stuffing-box forms the only joint requiring attention to keep it staunch, and it is accessible without removing the plunger. A piston is equivalent to a short plunger entirely contained within the cylinder and guided by it. The force is transmitted through a piston rod of relatively small area. Hence the piston has two faces on which the fluid pressure can act, and it is usually double-acting. With a piston there are two joints requiring to be kept staunch, one within the cylinder and one where the rod passes through the cylinder cover. 'A large hollow piston rod is termed a 'trunk.' The pistons of pumps are often termed 'buckets.'

85. *The volume swept through* by an ordinary piston is the product of the transverse section of the piston normal to the direction of motion and the length of its path. With an incompressible fluid, such as water, the volume swept through

is the volume of water lifted, in the case of a pump; or acting on the machine, in the case of a pressure engine or ram.

Work done on a piston.—The work done on a piston by fluid pressure is the product of the volume swept through by the piston and the intensity of the fluid pressure. If the fluid pressure is variable, the mean intensity of the fluid pressure is to be taken. If the work is to be in foot lbs., the volume swept through may be in cub. feet and the pressure in lbs. per sq. foot, or the volume swept through in units of 12 cub. inches and the pressure in lbs. per sq. in.

Velocity of piston.—Ordinarily a piston drives or is driven by a crank, rotating with nearly uniform velocity. Then the motion of the piston is approximate harmonic motion varying from rest at each end of the stroke to a maximum near mid-stroke. The acceleration is greatest at the beginning of the stroke, vanishes near mid-stroke, and changes sign and increases to another maximum at the end of the stroke.

86. *Influence of the weight of the piston on the crank-pin pressure.*—When a piston is driven by a constant pressure, it is generally desirable to make the piston as light as possible, because the inertia of the piston causes the piston effort to be more irregular than it otherwise would be. When, however, the pressure on the piston varies, the inertia of the piston may be used to diminish the variation of the piston effort and to make the total pressure on the crank pin nearly uniform.

Usually, when the inertia of the reciprocating parts is intended to equalise the effort on the crank pin in expansive engines, the weight, w, of the piston, piston rod, and cross head, reckoned per sq. in. of piston area, is so adjusted that at the intended speed of the engine $wv^2/g\text{R} =$ about $p/2$, where v is the velocity of the crank pin, R the radius of the crank, and p the initial steam pressure. (Units, feet and lbs.)

87. Construction of piston.—Various arrangements have been adopted to diminish the leakage between the piston and the sides of the cylinder in which it slides. The piston may be simply turned to fit the cylinder accurately (A, fig. 82); but, however good the fit at first, the wear of the cylinder and piston will gradually enlarge the clearance between them, and the leakage will steadily increase. If a series of recesses are cut round the piston circumference (B, fig. 82), the leakage for any given width of clearance

Fig. 82

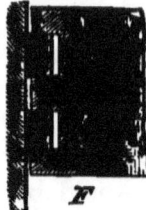

space is less, because the fluid loses its energy of motion, at each sudden enlargement of the section of the annular space between the piston and cylinder, through which it is escaping. Pistons of this kind are used for quick-running pumps, where a small leakage is not very prejudicial. Leakage may be prevented by placing, in a recess in the piston, a packing of gasket or tallowed rope (C, fig. 82). This soft and elastic packing is compressed against the cylinder by a junk ring, shown at *a*, which is fixed by studs or set screws. As the gasket wears away it can be replaced, and thus the permanent staunchness of the piston is secured. Pistons of this kind are not now much used for steam cylinders, though they are still employed for air pumps and cold-water pumps. The objection to them is that the repacking of the piston is troublesome, and the friction of the piston is considerable; also with high-pressure steam hemp packing is charred. To diminish the wearing away of the gasket, a face ring or spring ring, shown at *b*, was introduced (D, fig. 82), made of cast iron and divided on one side, to allow it to expand to the cylinder diameter as it wore away. The space behind the spring ring was at first filled by gasket packing, but it was found better to substitute steel springs for gasket, which retain their elasticity much longer, and press the spring ring outwards quite as effectively. In small pistons the elasticity of the spring ring itself is sufficient to maintain contact with the cylinder. The spring ring is then free from the piston. Various arrangements of this kind have been used. Sometimes the spring ring is a cast-iron ring, of uniform or varying thickness. Ramsbottom's rings are shown at E, fig. 82. These consist of a continuous spiral steel ring of three coils, or much more commonly of three separate steel rings, each split on one side. The rings are initially of one-tenth larger diameter than the cylinder, and, when compressed within it, press outwards with sufficient force to prevent leakage. The width of the rings (parallel to the axis of the cylinder) is about $0·014 D + 0·08$, and the thickness in the plane of the piston $0·025 D$, where D is the diameter of the cylinder. Ramsbottom's rings are usually of steel. It is stated that Mr. Ramsbottom found such rings tight against 100 lbs. steam pressure, when the radial pressure of the rings was $3\frac{1}{3}$ lbs. per sq. in. of surface. The rings were bent before being placed in the cylinder to a curve determined experimentally. A turned circular split ring of the required section was strained by weights acting on strings passing over pulleys, the total amount of the weights being the total amount of the required radial pressure of the ring. The curve was marked and the ring bent to the curve. Then, when compressed to its original circular form in the cylinder, it exerted a uniform radial pressure. Cast-iron rings answer very well for small spring rings. They retain their elasticity till half worn through. Cast-iron rings are sometimes of uniform thickness, but very often they are one-half thicker at the middle of the ring than at the ends where the ring is split. At F, fig. 82, is shown an arrangement for admitting the steam pressure in the cylinder to the back of the rings. In principle this is a good arrangement, but in this form it does not succeed very well, and is not very often adopted. In Stroudley's piston the steam is admitted to the back of the spring ring on the opposite side to that on which the steam acts. This ring, passing a little beyond the bored part of the cylinder, prevents the formation of a shoulder at the end of the cylinder. Bramah's cup leather is a perfectly successful application of the same principle. For pumps and blowing cylinders, wood blocks have been used to replace the spring ring. The packing of a piston may or may not share with the piston rim the pressure of the piston against the cylinder sides, whether due to the weight of the piston, as in horizontal engines, or to other causes.

The spring rings or metallic packing of pistons may be of cast iron, of wrought iron, of steel, or of gun-metal. For steam cylinders cast iron wears better than wrought iron, and about as well as gun-metal. Gun-metal is chiefly employed in pumps and in pistons of complicated types, the action of which would be impaired by corrosion. Steel is a good material for packing, especially where considerable elasticity is necessary. Cast-iron and wrought-iron rings may be made more elastic by hammering.

88. *Strength of pistons.*—Pistons are of a complicated form, and it is not easy to determine their strength theoretically. If the piston were a simple metal disc, supported at the centre and uniformly loaded, the greatest stress would be

$$f = k \frac{D^2}{t^2} p \quad . \quad . \quad . \quad . \quad (1)$$

where D is the diameter of the piston, t its thickness, p the greatest difference of the pressures on the two sides,[1] estimated per unit of area, and k is a constant. Putting $f =$ 8,000 for wrought iron and 3,000 for cast iron, we get

$$\left. \begin{array}{l} t = \cdot 0051 \, \text{D} \sqrt{p} \text{ for wrought iron} \\ = \cdot 0083 \, \text{D} \sqrt{p} \text{ for cast iron} \end{array} \right\} \quad . \quad (2)$$

These values of t will be taken as empirical units for

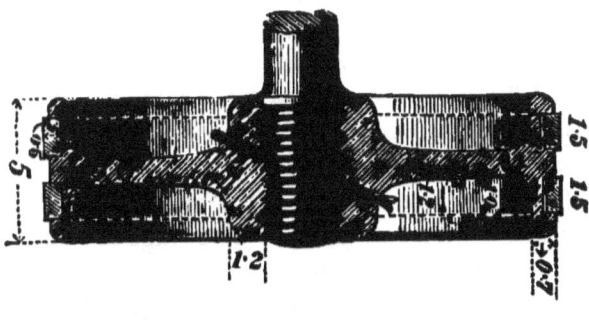

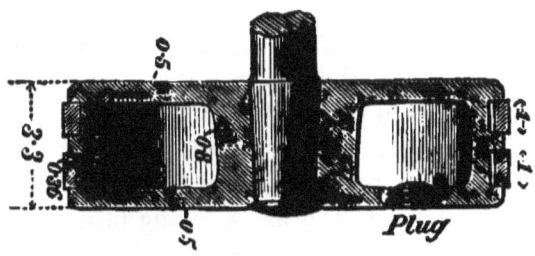

Fig. 83

the proportions of pistons. Since, however, the form of pistons varies greatly, and also the conditions under which they work, the draughtsman should not depend solely on

[1] That is, p is very nearly the initial absolute steam-pressure in the case of a condensing engine, and the initial steam-gauge pressure in a non-condensing engine.

the following proportional figures, but should deduce the proportional figures for himself, from good examples of pistons of a similar kind to the one he is designing.

89. *Locomotive pistons.*—Fig. 83 shows two forms of piston used in locomotives. One is constructed chiefly of cast, the other chiefly of wrought iron. Wrought iron is preferred by some engineers on account of its toughness and strength. But cast iron is much cheaper and answers well. Steel castings are now often used and are much lighter than cast iron. The spring rings in both cases are of cast iron and require no springs or packing. These rings are of uniform section, about $1\frac{1}{2}$ inch wide by $\frac{1}{2}$ inch thick, in pistons of average size. The split is made with a half lap, to prevent leakage at that point. The rings are sprung into the recesses in the piston, and should be so placed that the splits in the two rings are on opposite sides of the piston. This equalises the wear of the cylinder. A small screw is sometimes used to prevent the rings turning round in the grooves. The piston rod is screwed into the wrought-iron piston and fixed by a split pin. In the case of the cast-iron piston the rod is slightly coned at the end, and, when in place, is riveted over. The holes filled by screw plugs are intended for the removal of the sand core after casting.

Fig. 84 shows another locomotive piston. In this three spiral springs are placed behind the spring ring, and assist the elasticity of the latter in keeping the piston tight. A brass tongue-piece prevents leakage at the joint in the spring ring. The piston rod has a strong taper to enable it to be easily removed, and it is secured by a screwed end and nut. The spiral springs are so placed as to prevent the body of the piston bearing on the bottom side of an horizontal cylinder.

90. *Stationary engine pistons.*—Fig. 85 shows one form of stationary engine piston. It is made of cast iron, with a junk ring to confine the metallic packing. The packing consists of three cast-iron rings of the sectional

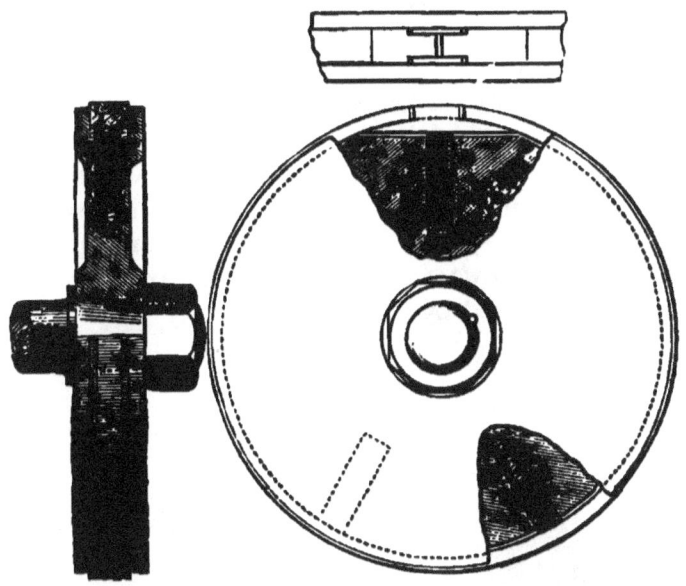

Fig. 84

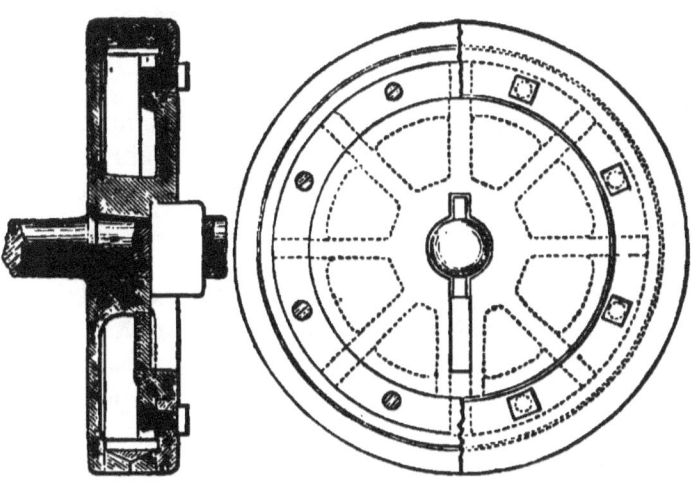

Fig. 85

form shown. The outer rings are turned $\frac{1}{16}$ inch larger than the cylinder diameter, and are split. The innner ring may or may not be split. By screwing down the junk ring the two outer rings are forced outwards, as they slide down the conical surfaces of the inner ring, and thus any desired amount of pressure can be obtained between the piston and cylinder. The inner ring has sometimes been made in the form of a spiral spring. It then presses the outer rings both apart and outwards. In this piston the rod is tapered at the end and fixed by a cotter.

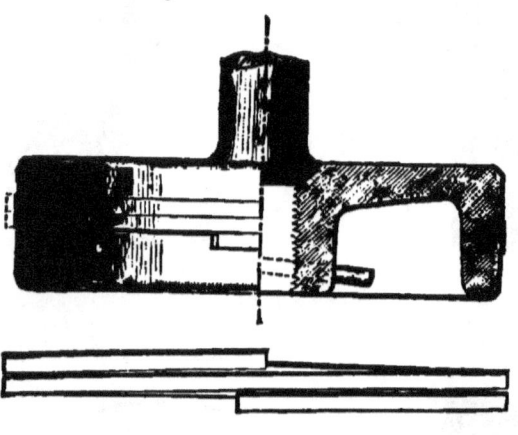

Fig. 86

Joy's piston (fig. 86).—The piston is a simple block, into which the piston rod is screwed and pinned. The diameter of this block is $\frac{1}{16}$ inch less than the bore of the cylinder. A recess is cut with a tool set to $\frac{1}{2}''$ pitch, and making 3 inches more than two revolutions. The cast-iron ring from which the packing is made is turned and bored, $\frac{5}{8}$ inch thick, and $\frac{3}{4}$ inch larger in diameter than the cylinder. It is then placed on a mandrel, and a spiral groove cut with a $\frac{1}{4}$-inch tool set at $\frac{5}{8}$-inch pitch, so as to form a spiral spring of $\frac{5}{8}$ inch by $\frac{1}{2}$ inch in section. The spiral spring gives probably a more uniform pressure on the cylinder than a simple ring, and its axial elasticity prevents its knocking in the groove.

Pistons and Piston Rods 145

91. *Marine-engine pistons.*—Marine-engine pistons are often of very large size, and are usually of cast iron, of a box shape and stiffened by numerous ribs. Fig. 87 shows

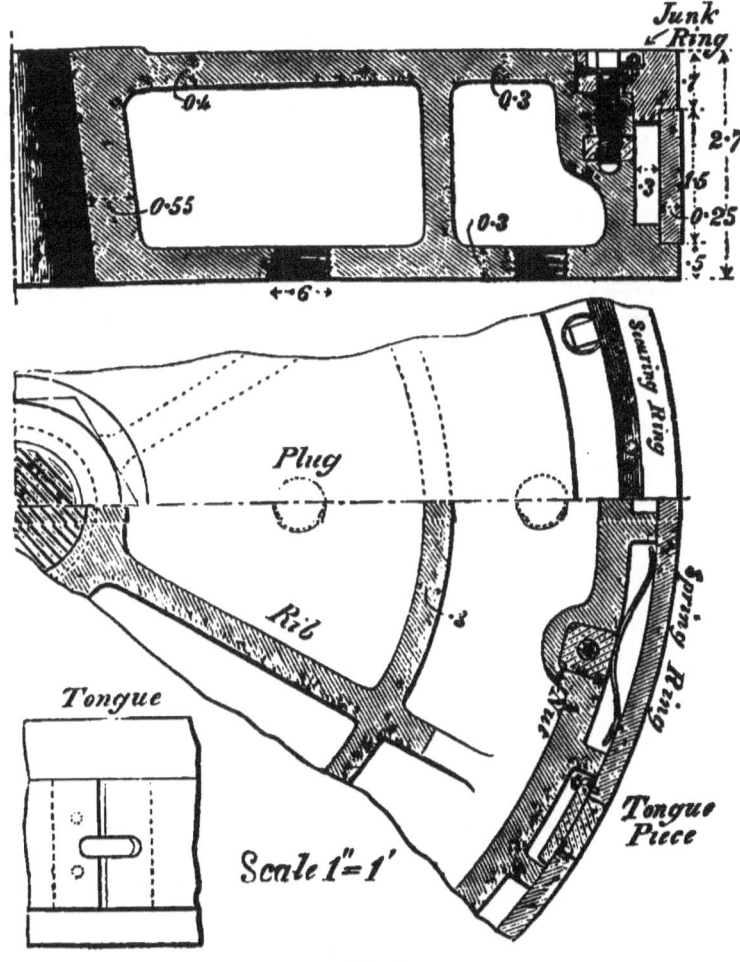

Fig. 87

a piston of this kind. The spring ring of cast iron is of uniform thickness. Leakage at the split is prevented by a brass tongue-piece, fixed to one end of the spring ring by screws. The split in the spring ring is shown square across the spring ring. But it is better to cut it obliquely so that the edges may not score the cylinder. The spring ring is pressed outwards by numerous plate-springs, placed in recesses cast in the rim of the piston. The spring ring and springs are kept in place by a junk ring. This last is attached to the piston by bolts, which have brass nuts placed in recesses behind the plate springs. To prevent these bolts slacking back, in consequence of the vibration of the piston, various locking arrangements are used. In the piston shown (a type used by Messrs. Humphreys and Tennant), a securing ring bears against the heads of all the junk-ring bolts. This ring is attached to the piston by studs, the nuts being fixed by split pins. Holes are cast in the piston for clearing out the loam core, and these are afterwards fitted with screw plugs or plugs secured by screwed pins.

In this piston the rod is tapered and the piston is secured to it by a large nut on the upper side of the piston. When

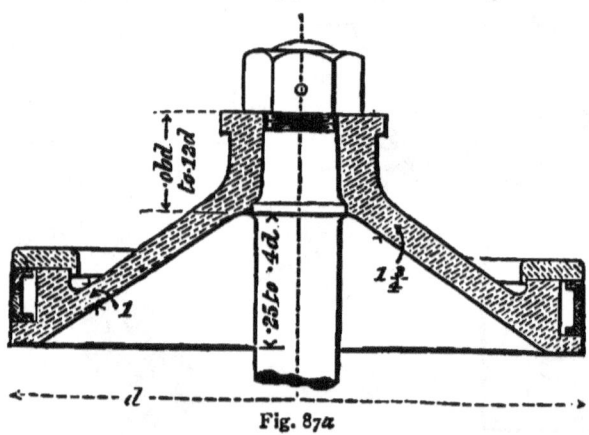

Fig. 87a.

such a piston works in a horizontal cylinder, a block placed between the spring ring and piston body on the bottom side of the piston keeps the latter from bearing on the cylinder. This block may extend round about one-fourth of the circumference.

Cast-steel pistons.—Large pistons are now often made of

steel castings, which are stronger and one-third lighter than cast-iron pistons. Fig. 87a shows a cast-steel piston such as is now adopted in marine engines. The unit for the proportional figures is $t = 0·0015d \sqrt{p}$ in large to $0·003d \sqrt{p}$ in small engines, but the thickness is in no case less than $\frac{3}{8}$ inch. The conical form gives strength and rigidity, and is at the same time easy to cast. The projection on the boss facilitates lifting the piston. The angle of cone is sometimes altered so that all the piston rods may be of the same size and length. The taper of the piston rod in the boss is 1 in 3 to 1 in 4, and the packing rings are pressed out by springs so that the pressure is about 3 to 5 lbs. per sq. in. of packing-ring surface.

92 *Hydraulic pistons.*—Fig. 88 shows a combined piston and plunger with a cup-leather or hat-leather arrangement for preventing leakage. The fluid pressure acting on the inside of the flexible leather-cups, aided by their own elasticity, makes an exceedingly staunch joint, whatever the pressure may be. The hat-leathers are so arranged that one acts when the piston moves in one direction, the other when the piston moves in the reverse direction.

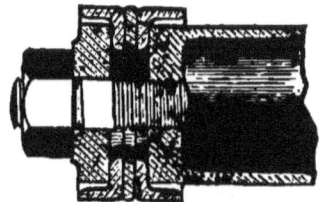

Fig. 88

The forms of leather packing for hydraulic cylinders will be described more fully in the next chapter.

THEORY OF SPRING RINGS FOR PISTONS

The object of the spring ring is to produce a uniform radial pressure sufficient to prevent leakage. It is known that under certain conditions the spring ring collapses, and then steam passes from the front to the back of the cylinder. The amount of radial pressure necessary to prevent this collapse must be determined experimentally. It is not at present very accurately known. It appears, however, that a radial pressure reckoned per square inch of surface of piston ring which is very considerably less than the steam pressure is sufficient to prevent this collapse of the piston ring. In any case it is desirable to determine what forms of spring ring secure a uniform radial pressure and what amount of radial pressure is to be expected from any given piston spring ring.

93. *Flexure by bending of a bar initially curved.*—In the case of a straight bar subjected to a bending moment, it is

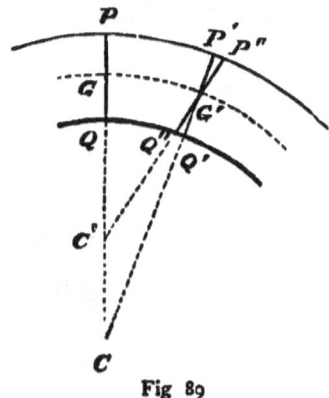

Fig 89

known (see I. § 27) that the bar takes a curvature given by the equation

$$M = \frac{EI}{\rho},$$

where M is the moment of the bending forces on one side of a section taken normal to the axis of the bar, I is the moment of inertia of the section, and ρ is the radius of curvature of the bent bar at the section.

Let fig. 89 represent a bar initially curved, and let G G' be a line passing through the centres of gravity of sections normal to G G'. Then if P Q, P' Q' are two normal sections at a small distance apart, and if P Q, P' Q' meet at C, C G = ρ is the radius of curvature at P Q of the unstrained bar. Now let a bending moment M be applied to the right of P Q. The particles initially in P' Q' will be found after bending in a new plane P'' Q'' still normal to the axis of the bar. The radius of curvature after bending will be $c' G = \rho'$. Then on the same assumptions made in treating straight bars, it will be found that

$$M = EI\left(\frac{1}{\rho'} - \frac{1}{\rho}\right)$$

that is, to pass from straight bars to curved bars, we have only to substitute in the equation for the curvature due to bending $\frac{1}{\rho}$, the augmentation of curvature $\frac{1}{\rho'} - \frac{1}{\rho}$.

94. *Theory of a cast-iron spring ring of unequal thickness.*—Let A B be a portion of a spring ring, which, initially circular on the outside, has been sprung into a cylinder of smaller diameter than itself. Let I be the moment of inertia of a cross section at B, which without sensible error may be taken nominal to the outside curve of the ring. Let ρ be the radius of the outside of the ring initially, and r its radius when sprung into the cylinder. Let b be the breadth, t the thickness of the ring at B. From the conditions in which the ring is placed, it is necessary to make the breadth b of the ring constant, and it is required to determine the variable thickness t of a ring which will produce a uniform pressure of p lbs. per sq. in. on the surface of the cylinder into which it is sprung.

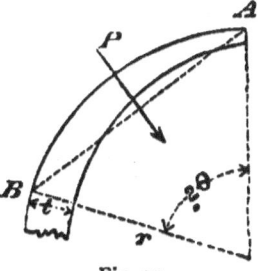

Fig. 90

The pressure of the ring on the cylinder per unit of length is $p\,b$, and the resultant of the uniform pressure from A to B is $p\,b \times$ chord A B, or

$$P = 2\,p\,b\,r \sin\theta.$$

The moment of this about B is

$$2\,p\,b\,r^2 \sin^2\theta.$$

Putting this value of the bending moment in the equation above,

$$2\,p\,b\,r^2 \sin^2\theta = EI\left(\frac{1}{r} - \frac{1}{\rho}\right)$$

But for a rectangular section, § 38,

$$I = \frac{b t^3}{12}$$

$$\frac{1}{r} - \frac{1}{\rho} = \frac{24 p r^2 \sin^2 \theta}{E t^3} \qquad . \qquad . \qquad (3)$$

For $2\theta = 180$, let t_1 be the thickness of the ring,

$$\frac{1}{r} - \frac{1}{\rho} = \frac{24 p r^2}{E t_1^3} \qquad . \qquad . \qquad (4)$$

Hence

$$\frac{t}{t_1} = \sqrt[3]{(\sin^2 \theta)} \qquad . \qquad . \qquad (5)$$

$2\theta =$	$t =$
10°	0·197 t_1
20°	·311
40°	·489
60°	·630
80°	·745
100°	·837
120°	·908
140°	·960
160°	·990
180°	1·000

From which table the thickness at different points in the ring can be calculated when t_1 is known.

Fig. 91 shows a ring drawn from these values. It is found that about ⅔ of the inside curve of the ring agrees very closely with a circle struck from a point c at a distance o c = a from the centre of the outside of the ring. It is easy to show that $a = 0·206\, t_1$ nearly. The ring can therefore be bored out to a radius $r - 0·794\, t_1$, and the points near A then thinned to agree with the proportions above.

From eq. 2,

$$p = \frac{E t_1^3}{24 r^2} \left(\frac{1}{r} - \frac{1}{\rho} \right).$$

Let ρ, the initial radius of the ring, be taken $1\cdot1\ r$, and put $E = 17{,}000{,}000$ for cast iron. Then

$$p = 64400 \frac{t_1^3}{r^3}.$$

Often in practice $t = \cdot06\ r$. Then

$$p = 14 \text{ lbs. per sq. in.};$$

a value which does not appear excessive as regards the wear of the cylinder.

It has been stated, however, that in experiments with Ramsbottom's rings, they have been found to be practically

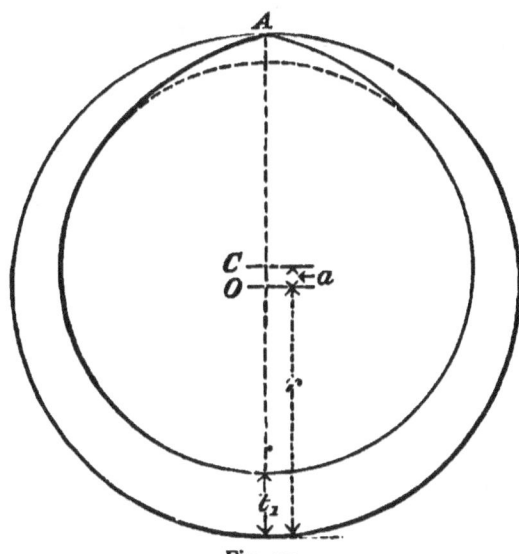

Fig. 91

steam tight when giving a uniform pressure of $3\frac{1}{2}$ lbs. per sq. in., even when the difference of pressure on the two sides of the piston amounted to 100 lbs. per sq. in. If we insert $p = 3\frac{1}{2}$ in the equation above, we get when $\rho = 1\cdot1\ r$

$$t_1 = 0\cdot04\ r,$$

a proportion which does not seem very exceptional.

Sometimes two eccentric rings are used placed one inside the other, as shown in fig. 92. The proportions of such rings given by Von Reiche are:

Total thickness of two rings $\delta = 0.8 + 0.12\, r$
Greatest thickness of outside ring $\tfrac{2}{3}\delta$
Least   ,,   ,, $\tfrac{1}{3}\delta + 1.2$
Greatest thickness of inside ring $\tfrac{2}{3}\delta - 1.2$
Least   ,,   ,, $\tfrac{1}{3}\delta$
Breadth of rings   .   .   . $0.12\, r + 0.9$

It is easy to see that such rings will produce approximately the same pressure on the cylinder as a single ring designed as described above, and of a maximum thickness equal to the total thickness of the two rings.

The breadth of the ring does not affect the intensity of pressure against the cylinder, nor for a given pressure per unit of area will a narrow ring wear longer than a wide one. On the other hand, a wide ring will wear the cylinder more than a narrow one.

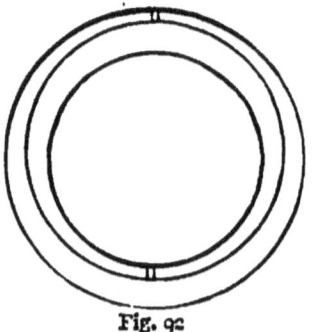

Fig. 92

The theory above is simplified from an investigation by Prof. Robinson,[1] which contains also an attempt to determine theoretically the proper pressure of the ring on the cylinder to prevent leakage. This latter part of the investigation does not appear to the author to be well based.

95. *Theory of a spring ring of uniform thickness.*—A spring ring of uniform thickness may be made to give a uniform pressure on the cylinder, if it is initially bent to a form of varying curvature. Let A B C, fig. 92, be the spring ring when sprung into the cylinder, A' B C the ring when unstrained. Let b be the breadth and t the thickness of the ring; I the moment of inertia of a cross section; ρ the radius of curvature of a point B when the ring is unstrained; r the radius of the cylinder into which the ring is sprung. The resultant pressure between A and B is as before, .

$$2\, p\, b\, r\, \sin \theta.$$

[1] Van Nostrand's Magazine, June 1881.

Its moment about B is
$$M = 2pbr^2 \sin^2 \theta.$$
Inserting this in the equation above,
$$2pbr^2 \sin^2 \theta = EI\left(\frac{1}{r} - \frac{1}{\rho}\right)$$
$$\frac{1}{r} - \frac{1}{\rho} = \frac{24pr^2 \sin^2 \theta}{Et^3}.$$

Hence
$$\rho = \frac{Et^3 r}{Et^3 - 24pr^3 \sin^2 \theta} \qquad . \qquad . \quad (6)$$

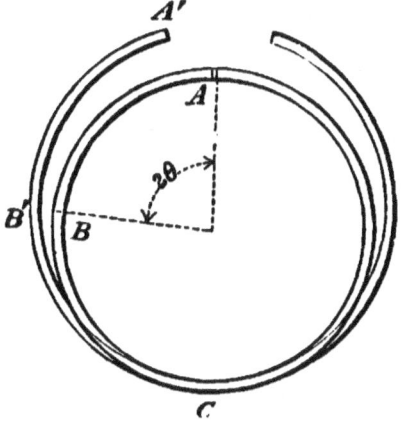

Fig. 93

Let $E = 30,000,000$ for steel, and let $t = 0.04\, r$, which is about the usual proportion in practice. Then
$$\rho = \frac{80\, r}{80 - p \sin^2 \theta}.$$
The following table gives values of ρ for $\theta = 0°$ to $90°$, or $2\theta = 0°$ to $180°$, and for $p = 3\frac{1}{2}$ and $p = 14$, the values assumed in the previous case.

$2\theta =$	$\rho =$ for $p = 3\frac{1}{2}$	$= \rho$ for $p = 14$
0°	1·000 r	1·000 r
30	1·003	1·011
60	1·012	1·047

90	1·021	1·096
120	1·033	1·152
150	1·042	1·195
180	1·045	1·214

The form to which the spring ring ought to be bent in order that when sprung into the cylinder the pressure may be uniform, may be obtained thus :

Let o be the centre of the cylinder. Divide its circumference at $a\,a'\,a''$.... into 12 equal parts. Take $ac=$ the

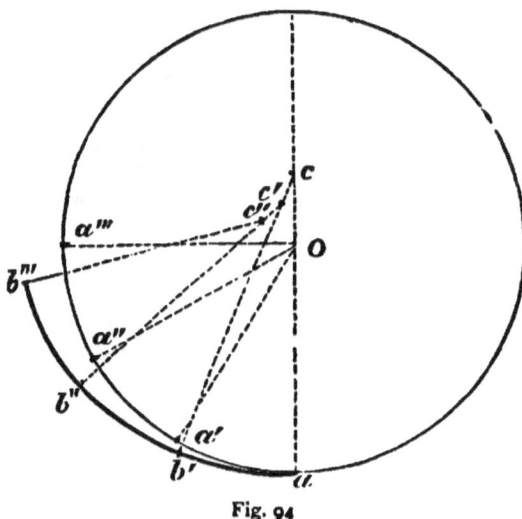

Fig. 94

value of ρ for $2\theta=180$, and with centre c draw the arc $a\,b'$. Make $a\,b'=a\,a'$ and join $b'\,c$. Take $b'\,c'=\rho$ for 150°, and with centre c' draw the arc $b'\,b''$. Proceeding thus, the whole curve can be drawn.

Fig. 95 shows the curves drawn for the values of ρ for $p=3\frac{1}{2}$ and $p=14$ given above. The former curve agrees roughly with a circle of radius $\frac{1}{25}$th greater than the cylinder radius; the latter curve with a circle of radius $\frac{1}{7}$th greater than the cylinder radius. But it will be seen that both curves deviate considerably from circular curves, and hence rings made by turning them to a circular form and cutting out a portion cannot give a nearly uniform distribution of pressure.

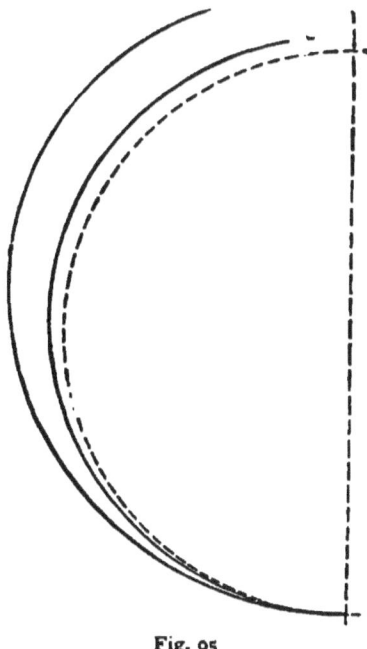

Fig. 95

PISTON RODS

96. Piston rods are subjected alternately to compression and tension. In horizontal engines of large size, the weight of the piston and rod produces bending, and extra strength is then required. In oscillating engines, the resistance of the cylinder to angular acceleration also causes bending of the piston rod. In most engines the piston has a single piston rod, but there are some types of engine (return connecting-rod engines and some forms of tandem compound engine, for instance) in which there are two piston rods. It is more difficult to fit two rods accurately, so that they shall be equally strained, and hence double-piston rods should have some extra margin of strength.

For the minimum section of the piston rod through the crosshead cotter hole, or at the bottom of the screw thread at the piston nut, seeing that the stress acts only in one direction, a stress might be allowed of 10,000 lbs. per sq. in. for wrought iron, or 13,200 for soft steel (Table II. Part I. p. 43). But it is probably impossible to adjust a cotter so that the stress is uniformly distributed, and hence a lower value of the mean working stress must be taken. Let D be the piston diameter in inches, p the greatest difference of pressure on the two sides in lbs. per sq. in., a the section of the rod after deducting cotter hole, m a factor of safety to allow for straining actions due to inertia, water in the cylinder and unequal distribution of stress, f the safe working stress on the material given above. Then the piston load due to steam pressure is $\frac{\pi}{4} D^2 p$, and the section of the rod is determined by the relation

$$f a = m \frac{\pi}{4} D^2 p.$$

It appears that m should not be taken less than 2, although cases will be found of rods working with a less margin of security. However, taking $m = 2$ and $f = 10,000$ for wrought iron, and 13,200 for steel—

$$a = 0{\cdot}000157 \, D^2 p \text{ for wrought iron,}$$
$$= 0{\cdot}000119 \, D^2 p \text{ for steel.}$$

If $d_{min.}$ is the minimum diameter of the piston rod, and the net section after deducting the cotter hole is taken to be $0{\cdot}535 \, d^2_{min.}$, then

$$d_{min.} = 0{\cdot}0171 \, D\sqrt{p} \text{ for wrought iron,}$$
$$= 0{\cdot}0149 \, D\sqrt{p} \text{ for steel.}$$

97. *Size of cotter for piston rod.*—For the piston load given above and using the rules in Part I. p. 175, for cotters strained alternately in opposite directions, we get

$$m \frac{\pi}{4} D^2 p = 2 \, b \, t f_s,$$

where b is the width, t the thickness of the cotter, and $f_s = 4{,}000$ for wrought iron, and 5,300 for steel. It appears that for the cotter the stress is pretty uniformly distributed, so that $m = 1\frac{1}{4}$. Hence

$$b t = 0\cdot 000122\ \mathrm{D}^2 p \text{ for wrought iron,}$$
$$= 0\cdot 000092\ \mathrm{D}^2 p \text{ for steel.}$$

It $b = 4 t$, a common proportion,

$$t = 0\cdot 00552\ \mathrm{D}\sqrt{p} \text{ for wrought iron,}$$
$$= 0\cdot 00479\ \mathrm{D}\sqrt{p} \text{ for steel.}$$

98. *Strength of piston rod considered as a strut.*—The rules above fix the minimum section of the piston rod, but the body of the rod has to be of greater section to secure stiffness and resistance to bending. The rod is sometimes treated as a long column, but this is unsatisfactory, partly because the precise condition of loading at the ends is unknown, and partly because there are transverse bending actions, due for instance to the slide bars being slightly out of line with the cylinder, which it is impossible to calculate. It is enough, therefore, to design the rod simply for a low value of the working stress, shown by experience to give margin enough of strength to resist unknown straining actions. Taking the load on the rod to be as above, $\frac{1}{4}\pi \mathrm{D}^2 p$, due to the steam pressure, then the following values of the working stress have been found in actual engines. In short stroke direct-acting engines $f = 3{,}600$ to 3,000; in return connecting-rod engines $f = 3{,}000$ to 2,400; in moderately long stroke horizontal engines $f = 2{,}500$ to 2,000; in oscillating engines $f = 2{,}000$. Hence, if d is the diameter of the rod, either of steel or iron,

$$d = k\ \mathrm{D}\sqrt{p}$$

$f =$	$k =$
2,000	0·0224
2,500	0·0200
3,000	0·0182
3,500	0·0169

158   *Machine Design*

Very often the minimum diameter of the rod, allowing for the cotter hole, is also sufficient for the diameter of the body of the rod. In compound engines for reasons of convenience the piston rods are often all of the same diameter. Then the rod must be determined for the piston which has the greatest load.

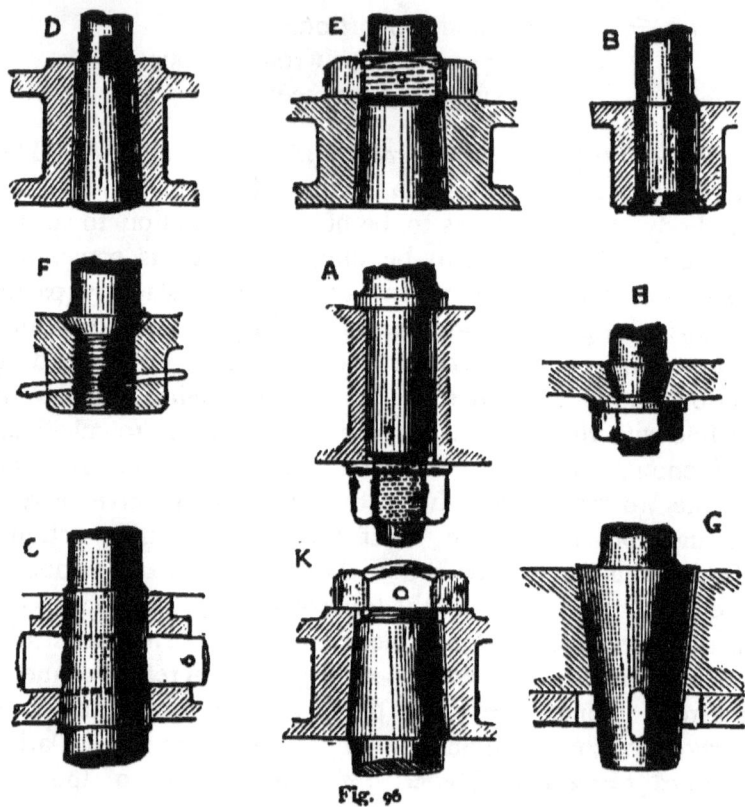

Fig. 96

99. *Modes of fixing piston to piston rod* —For pump buckets and other cases where a very accurate fit is not required the rod may have a shoulder, a parallel part in the piston and a nut at the back of the piston (fig. 96, A). With steam pistons, which must be steam-tight, in some cases the rod is turned slightly larger than the hole in the piston. Then

the piston is heated and shrunk on, and the rod riveted over (fig. 96, B). The diameter of rod may be made 1·0025, the diameter of the hole bored in the piston. Usually, however, it is necessary that the piston should be easily removed from the piston rod. Then the part of the rod in the piston is in part or wholly tapered. If the taper extends through the piston the taper is 1 in 32 to 1 in 16 (that is, the total taper for the two sides). Sometimes the rod has a short tapered part with a taper of 1 in 4, till the section of the rod is reduced to three-fourths its full section and the remainder is parallel and not tightly fitted. In Mr. Stroudley's piston, F, there is a still sharper taper, the rod being enlarged to permit it.

CHAPTER VII

STUFFING-BOXES

100. Stuffing-boxes are used to prevent leakage of steam or water at the points where moving parts pass through the sides of vessels containing fluids. Thus a stuffing-box is used where the piston rod of an engine passes through the cylinder cover, or where a rotating shaft passes through a centrifugal pump case. In ordinary stuffing-boxes, soft packing is used to prevent leakage, but various forms of metallic packing have also been employed.

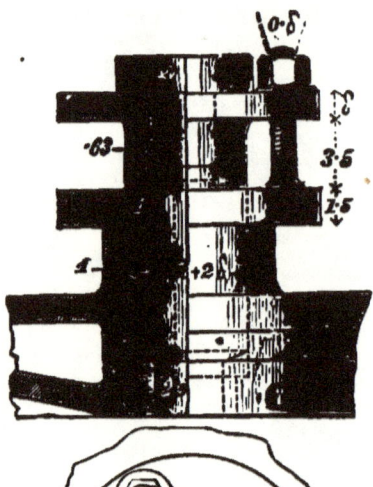

In fig. 97 is shown a stuffing-box for a vertical rod, and in fig. 98 a stuffing-box for a horizontal rod. In both cases the stuffing-box is cast on the cylinder cover. The stuffing-box is larger than the rod which traverses it, by the space necessary for the

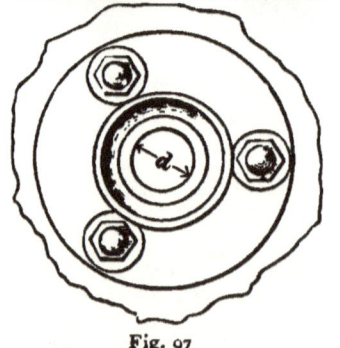

Fig. 97

soft packing. At the bottom of the box there is a brass bush, which, being softer than the rod, preserves the latter from injury. When the bush has worn oval, it is easily replaced by a new one. To keep the packing in place and to compress it sufficiently to prevent leakage, a loose piece termed a 'gland' is used. This is entirely of brass (fig. 98) or bushed with brass (fig. 97), and often has an oil-box formed in it (fig. 98). The gland is forced down on the soft packing by two or more bolts or studs.

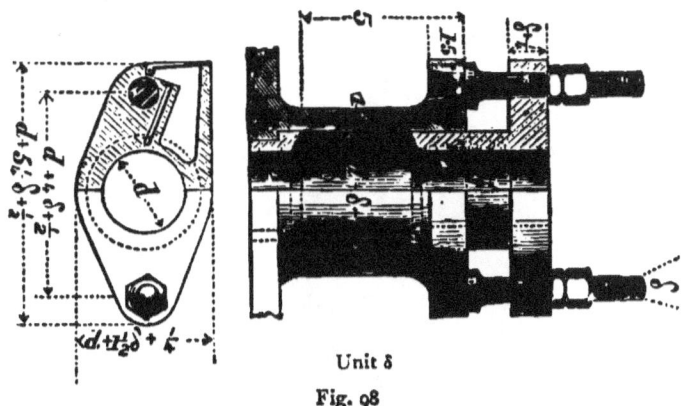

Fig. 98

101. *Proportions of stuffing-box and gland.*—Let d be the diameter of the rod or shaft traversing the stuffing-box. The diameter δ of the gland bolts may be $= \frac{1}{4}d + \frac{1}{4}$, if there are two; and $= \frac{1}{5}d + \frac{1}{4}$, if there are three. Then δ will be taken as the unit for the proportions of the box. The thickness of packing in the box is very variable. In ordinary stuffing-boxes, not of very great diameter, the packing thickness varies from $\frac{1}{2}\delta$ to δ. But in large stuffing-boxes for trunks or hollow piston rods a less thickness is employed. The length of the box is also variable. The greater the length of box, the less frequently will it be necessary to renew the packing. On the other hand, the space available for the stuffing-box is sometimes restricted. From 5δ to 8δ is an average length. The thickness of the stuffing-box flange may be $1\frac{1}{4}\delta$ to $1\frac{1}{2}\delta$, and the thickness of the gland flange may be δ for cast iron, or $1\frac{1}{4}\delta$ for brass. If an oil-box is cast in the flange, the thickness is somewhat greater. The length of gland may be $\tfrac{3}{4}$ to $\tfrac{7}{8}$ the stuffing-box length. The thickness of the stuffing-box should not be less than $\tfrac{1}{5}\sqrt{d}$ or less than $\tfrac{3}{8}\,\hat{c}$. The length of the brass bush may be about $2\tfrac{1}{2}\,\hat{c}$, but when the stuffing-box serves to guide the rod, as in oscillating engines, a much greater length of bush is used.

102. *Yarrow's stuffing-box.*—As ordinarily constructed the gland of a stuffing-box fits the piston rod, and at the same

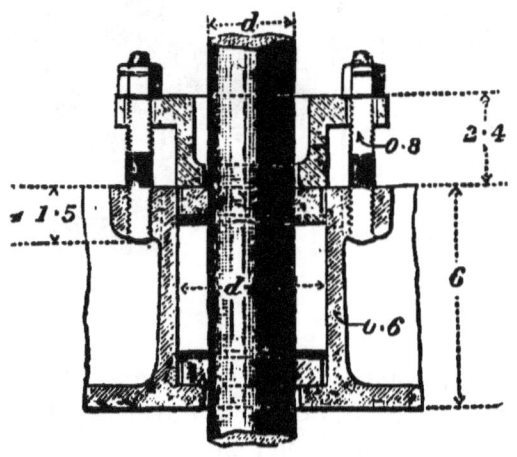

Unit $\tfrac{1}{4}\,d+\tfrac{1}{4}$

Fig. 99

time it is fixed in position by the stuffing-box. Hence it must act as a guide to the piston rod. The gland ought, therefore, to be exactly coaxial with the cylinder and parallel to the line of stroke of the slide block. If initially or in consequence of wear these conditions are not satisfied, the piston rod must bear heavily on one side of the gland, and will wear it oval. In quick-running engines the piston rod will be heated on the side bearing on the gland, and this will cause it to bend in a direction which increases the injurious pressure. Messrs. Yarrow have found it advantageous to guide the parts connected by the piston rod by the piston and slide block only, and to construct the stuffing-box so as to permit a small lateral adjustment of position of the piston rod without causing pressure and heating. Fig. 99 shows this stuffing box. The gland is bored out a little larger than the piston rod (about $\frac{1}{8}''$ larger) and the ordinary fixed bush at the bottom of the stuffing-box is dispensed with. Two

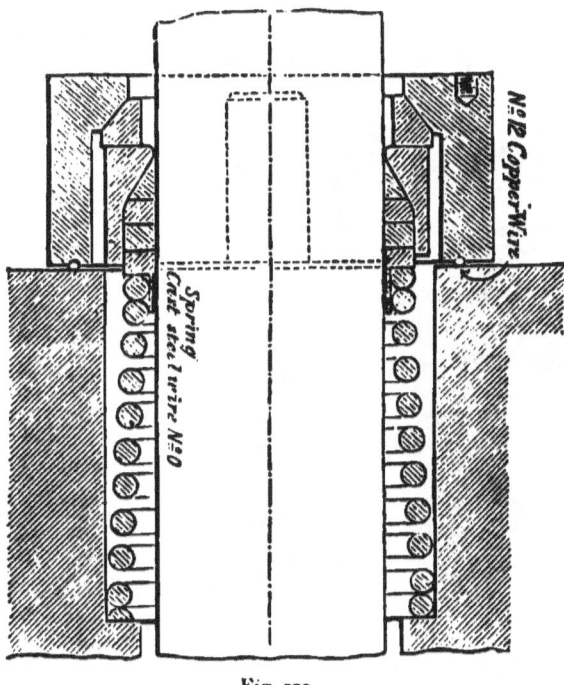

Fig 100

rings fit closely to the piston rod, but are turned a little smaller than the stuffing-box, and between these and the soft packing are two thin washers fitting the stuffing-box, but $\frac{1}{8}''$ smaller than the rod.

103. *Stuffing-boxes with metallic packing.*—Packing consisting of metallic rings in place of the ordinary soft packing has often been tried with more or less success. It should

require much less attention to keep in order than soft packing. Fig. 100 shows one good form for piston-rod stuffing-boxes. Three Babbit metal split rings grasp the piston rod. These fit inside a cup which presses on a ring having a spherical surface fitting the fixed part of the stuffing-box. Hence the piston rod can move a little sideways by displacing the cup and angularly by displacing the ring in its spherical seating, without causing leakage. The flat and spherical joints between the cup and ring, and between the ring and seating, are carefully ground, so as to be steam tight in any position. The Babbit metal rings are kept in place in the cup by a spiral spring acting on a distance piece which is inserted a little way into the cup. The pressure of the steam acts with the spring in keeping the rings in place. The essential feature is that steam tightness is obtained without hindering the unavoidable lateral movement of the piston rod. The wear of the piston rod with this packing is very small. No doubt this form of packing is greatly superior to soft packing. It requires, however, the best workmanship, and is necessarily expensive in first cost.

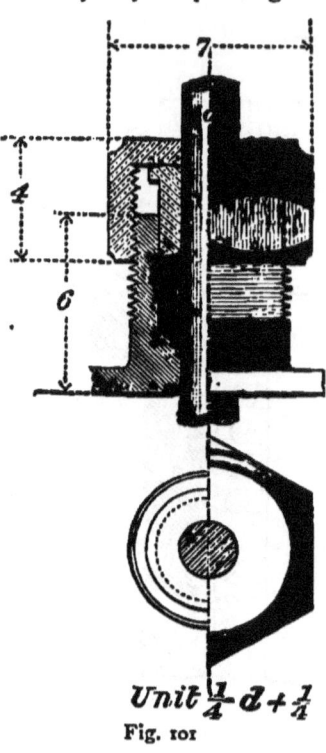

Unit $\frac{1}{4}d + \frac{1}{4}$

Fig. 101

104. Small stuffing-box.—For small rods, such as those round the spindles of valves and cocks, the form of stuffing-box shown in fig. 101 is used. The stuffing-box has a screw thread on the outside, and a six-sided cap fits over the gland and is screwed to fit the stuffing-box.

Taking the unit as $\delta = \frac{1}{4}d + \frac{1}{4}$, the internal diameter may be $2\frac{1}{2}\delta$, the external diameter $5\frac{1}{2}\delta$, and the other proportions as given in figure.

105. *Packing for stuffing-boxes.*—A loose kind of hemp rope termed 'spun yarn,' steeped in melted tallow, was commonly used for stuffing-box packings. Occasionally brass turnings were sprinkled over the spun yarn, to cause the packing to wear longer. Packings of india-rubber wrapped in canvas, termed 'elastic core packing,' and metallic packings of woven wire, are also used. These are formed into ropes of such a form that when cut of a length equal to the circumference of the rod, they form a ring exactly fitting the stuffing-box. Lately asbestos prepared in discs or ropes has been used. Elastic core or asbestos packing is much better than spun yarn for engines working with high-pressure steam.

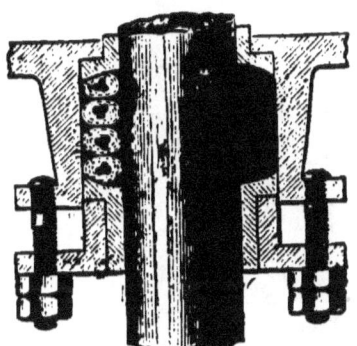

Fig. 102

Fig. 102 shows the most common mode of using a coil of india-rubber or asbestos packing in stuffing-boxes. The advantage of asbestos is that it stands the high temperature of steam at the high pressures now used without injury.

Glands when worn, if not bushed at first, are bored out, fitted with a bush having a collar at the inner end, and the bush is then bored.

For hydraulic purposes, Mr. Tweddell has found hemp packing in ordinary stuffing-boxes efficient under pressures of 1,500 to 2,000 lbs. per sq. in.; and where the rod passing through the stuffing-box is continuously at work, the hemp packing gives less trouble in such cases than a cup-leather.

106. *Cup-leather packing.*—When great hydraulic pres-

sure is to be resisted, a peculiar packing is used, invented by Bramah, and already alluded to (§ 92). The leakage is prevented by a flexible leather ring, kept in contact with the piston rod or ram on one side and the cylinder on the other by the fluid pressure. In its original form, leather was moulded into an annular shape in plan and to a U-shape in section (fig. 103). This ring is placed in a recess in the cylinder or in a stuffing-box, in such a way that the fluid has free access to its interior; the fluid pressure acting within the ring presses it against the plunger and the sides of the recess, and this, aided by the elasticity of the ring, makes a perfectly

Fig. 103

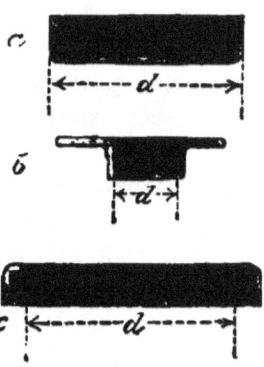

Fig. 104

tight joint. When the cup-leather is large, it is provided with an internal brass ring. At one time a thin guard ring of brass was used on the edge most liable to wear. Mr. Tweddell, who introduced this guard ring, no longer uses it. A packing of hemp or cotton is sometimes used as a bed for the leather.

Fig. 104 shows all the ordinary forms of hydraulic leather packing now used. *a* is a cup-shaped leather used for pump buckets; *b* is a form sometimes termed a 'hat' leather; *c* is the ordinary Bramah cup-leather or double cup-leather.

Stuffing-boxes

Mr. Welch gives the following rules for the proportions of cup-leathers for great pressures. Let D be the diameter of the ram or plunger. Then the thickness of the leather should be

$$t = 0.156\, D^{0.28} \qquad . \qquad . \qquad . \quad (1)$$

$$\log. t = \frac{7}{25} \log. D - 0.8069,$$

and the width and depth of the ring measured outside should be each $2\frac{1}{2} t$. ('Proc. Inst. Mech. Eng.' 1876.)

D =	3	6	9	12	15	18	21	24
t =	0·212	0·258	0·288	0·313	0·333	0·351	0·366	0·380

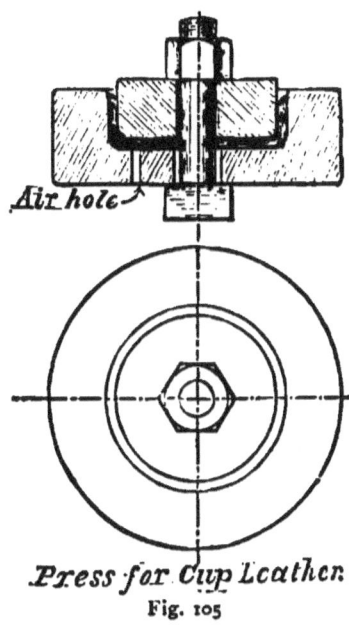

Press for Cup Leather.
Fig. 105

Fig. 105 shows a press for moulding cup-leathers.[1] The best oil-dressed leather is steeped in warm water, and then forced gradually into the mould and left till it is again hard.

[1] Anderson, 'Chatham Lectures on Hydraulic Machinery.'

Friction of cup-leathers.—According to experiments by Mr. John Hick, the friction of a press cup-leather on a ram D inches in diameter with a pressure of p lbs. per sq. in. is $F = c D p$, where c is a constant. But the whole pressure on the ram is $\frac{1}{4}\pi D^2 p$. Hence the fraction of the ram pressure expended in friction is $4c/\pi D$. According to the experiments of Mr. Hick, c has values from 0·03 to 0·05 when p is in lbs. per sq. in. and D is in inches. Some experiments on cup-leather friction were given in 'Engineering,' June 15, 1888, and in these it appeared that the friction was 4 to 9 per cent. of the ram pressure for leathers in good condition. In one case the friction amounted to 19 per cent. of the ram pressure.

CHAPTER VIII

FLYWHEELS

107. For most of the purposes for which engines are used, considerable regularity of speed is important. Now there is not in general an equilibrium between the effort exerted by the engine and the resistance overcome. Both the effort and resistance vary, and though for any sufficiently long period the mean effort and the mean resistance must be equal, there is during the period an alternate excess of effort or of resistance, producing a fluctuation of speed. It is to moderate this fluctuation of speed that heavy flywheels are used, which alternately store and restore a portion of the energy of the engine.

The causes of these temporary fluctuations of speed are chiefly: (*a*) the variation of the effective steam pressure driving the piston; (*b*) the variation in rotative engines of the leverage at which the piston pressure acts in rotating the crank; (*c*) periodical variations in the resistance overcome, such as the variation, during a stroke, of the pressure on a pump driven by the engine. We have not here to do with permanent causes of alteration of speed such as variation of boiler pressure, or permanent alteration of work being done by throwing machines into or out of gear.

108. *Flywheel radius and speed.*—The energy of motion of a steam engine at any given moment is partly the energy of reciprocating pieces, partly the energy of revolving pieces. The reciprocating pieces have variations of speed definitely connected with the piston positions, and since the whole

energy stored in, or restored by, these is not a large fraction of the whole energy of motion, we may neglect the small differences of speed of these in different strokes. Treating the reciprocating pieces as always having the same speed at the same points of the stroke, the velocity being calculated as if the crank pin revolved uniformly, the forces producing changes of energy of motion in the reciprocating pieces can be dealt with, without appreciable error, as additions to or deductions from the steam pressure on the piston. If the indicator diagram is corrected for the inertia of the reciprocating pieces, § 44, then the influence of these on the fluctuations of speed of the engine need not further be considered.

Of the rotating pieces there is one, the flywheel, of such very large weight and moment of inertia, that the effect of all the others on the fluctuations of speed may be disregarded. Let w_r be the weight of the rim, w_a the weight of the arms and nave, ρ_r the radius of gyration of the rim, and ρ_a the radius of gyration of the arms. Let ΔE be a quantity of energy stored or restored during a change of angular velocity from ω_1 to ω_2. Then

$$\Delta E = \left(W_r \rho_r^2 + W_a \rho_a^2\right) \frac{\omega_2^2 - \omega_1^2}{2g}.$$

If the rim is of rectangular section and internal and external radii R_1 and R_2, then

$$\rho_r^2 = \frac{R_1^2 + R_2^2}{2}.$$

If R is the radius to the centre of figure of the rim, $R = \frac{1}{2}(R_1 + R_2)$, and ρ_r will not in ordinary cases differ much from R. Also as the weight of the arms is a good deal less than that of the rim, and ρ_a does not greatly differ from $0.577 R$, we may take for practical purposes

$$\Delta E = (W_r + \tfrac{1}{3} W_a) R^2 \frac{\omega_2^2 - \omega_1^2}{2g}. \qquad (1)$$

If $v_1 v_2$ are the velocities at the radius R, this becomes

$$\Delta \mathrm{E} = (\mathrm{W_r} + \tfrac{1}{3}\mathrm{W_a}) \frac{v_2^2 - v_1^2}{2g} \quad . \quad . \quad (1a)$$

Hence, to reduce the fluctuation of speed for any given excess or defect of energy, $\Delta \mathrm{E}$, as much as possible, it is necessary to put as much weight as possible into the rim of the flywheel, and to make its radius as large as possible.

In the older steam engines the flywheel radius was generally 1·7 time the stroke. It is now sometimes 2 to $2\tfrac{1}{2}$ times the stroke. But there are limits both of convenience and safety which restrict the increase of flywheel radius. The centrifugal force, due to rotation, produces a centrifugal tension in the rim, and if the peripheral velocity exceeds certain limits the wheel becomes unsafe. Of all machinery accidents, the bursting of flywheels, or, as such accidents are termed in Germany, the 'explosion of flywheels,' is among the most destructive. Anyone who will consult the list of such accidents given in a paper by Herr Köchy in the 'Verhand. des Vereins zur Beförderung des Gewerbfleisses' for 1886, will see that such accidents are not infrequent.

Let N be the revolutions per minute of the flywheel, reckoned at the centre of the rim, then $v = \pi \mathrm{R} \mathrm{N}/30$ is the velocity of the rim in feet per second. For pulleys and wheels with solid rims, that is, rims without any joints, a peripheral speed $v = 100$ feet per second is about the limit found safe in practice. But when flywheels exceed 10 feet in diameter it is generally necessary to construct them of segments bolted together, and then so high a velocity is hardly safe. For a long time it has been usual in this country to use a spur flywheel for factory engines working on to a mortice spur pinion on the first motion shaft. The necessary speed for the shafting is thus gained in the most simple and direct way. Such spur flywheels are most commonly built of segments, and for safety the limiting speed has very generally been taken at 30 to 40 feet per second. Some further particulars of the speed of such wheels will be found in Part I. p. 288. It is worth noting, however, that these usual speeds are often exceeded with safety. Radinger, for instance ('Proc. Inst. Civil Engineers,' lv. 404), has given particulars of some American flywheels, in which these limits of speed are greatly exceeded:

	Corliss Engine	Amoskeag Mfg. Co.	Bay State Mill
H.P. transmitted	1,400	570	400
Diameter, feet	30	28	20·5
Revs. per min.	36	49	48
Velocity of pitch line, feet per second	56·5	72	51·5
Pitch	5·23	5·18	4·02
Breadth	24	18	18
Pressure per inch of breadth	567	242	237

Radinger attributes the high values of speed and pressure found in these wheels to their superior workmanship; obviously in such cases the strength of the joints of the arms and rim should be very carefully studied. There is one case, however, in which even these speeds are exceeded. Flywheels are commonly used for engines driving rolling mills, which store an amount of energy when the rolls are running empty, which is afterwards restored in forcing the heated bar through the rolls. Such wheels are exposed to excessively severe straining action in consequence of the great and almost sudden changes of velocity they undergo. Nevertheless, they are not rarely run at speeds of 90 feet per second or even 100 feet per second. With such wheels, however, it is also true that most of the accidents occur.

Mr. Halpin has proposed to construct a flywheel rim of great strength by putting a worm and wheel to drive the crank shaft, and winding on to the rim steel wire of great tenacity. A flywheel of this type has actually been used for the engines driving the machinery for rolling tubes by the Mannesmann process, where an enormous power is required for a very short time.

Determination of Weight of Flywheel for Given Fluctuation of Speed

109. *Determination of the coefficient of fluctuation of energy.*—The first problem to be dealt with is the determination of the weight of flywheel necessary to secure a limitation of the fluctuation of speed within assigned limits. No doubt very rough rules of thumb are commonly used in assigning the flywheel weight in many cases, and these insure a practically sufficient regularity. But they do not secure that the flywheel has only sufficient weight for the purpose in view, and excessive flywheel weight not only causes loss of work in friction, but in some cases, such as engines driving large pumps, may be dangerous. Two cases of practical importance may be distinguished : 1. A rotative engine drives a number of machines. In this case the variations of resistance of the different machines in any short period sensibly balance each other. The resistance overcome by the engine may be treated as practically constant, and it is only necessary to estimate the excess and deficiency of the energy of the engine during a semi-revolution. 2. An engine drives a pump or air compressor connected directly with its piston rod. In that case the fluctuation of resistance as well as the fluctuation of effort in each semi-revolution must be taken into account. The case of an engine driving a single machine with a known variation of resistance, but by gearing so that it is only in a series of revolutions that there is an exact balance between motive and resisting work, is too complex to be treated here.

In Chapter III. methods have been given for finding the crank-pin effort very exactly from the indicator diagram, and these may be used in dealing with flywheels. But it will be seen presently that some approximations are necessary in treating flywheels to avoid too great complexity. Hence it will be quite accurate enough to neglect the effect of the obliquity of the connecting rod. Then the determination of the crank-pin effort is very simple.

110. *Periodical excess and deficiency of energy in a rotative engine working against a practically constant resistance.*—Let A B (fig. 106) represent to any scale the stroke of the engine, the semicircle being the path of the crank pin in a single stroke. On A B draw a diagram A *a b c* B the ordinates of which are the effective steam pressures on the piston; that is, the ordinates are the intercepts between the forward pressure line of an indicator diagram and the back pressure line for the other end of the cylinder (§ 47). Neglecting the connecting-rod obliquity, for any position 2 of the crank pin, *h* is the position of the piston, and *h k* the effective piston effort in lbs. per sq. in. Before proceeding further it is desirable to correct this diagram for the inertia of the reciprocating parts.

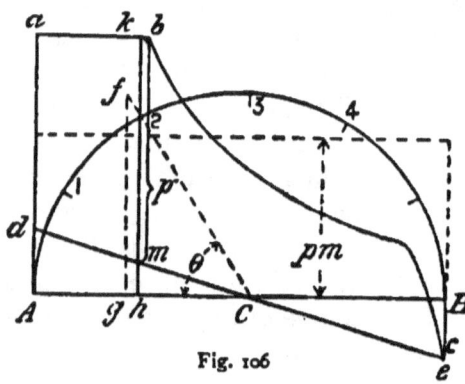

Fig. 106

Take A*d* = B*e* = the accelerating force of the reciprocating pieces at the ends of the stroke reckoned in lbs. per sq. in. of piston. It has already been shown (§ 42) that if *w* is the weight of reciprocating parts per sq. in. of piston, *v* is the mean crank-pin velocity, R the crank radius,

$$A d = B e = \pm \frac{w}{g} \cdot \frac{v^2}{R}.$$

Join *d e*. Then the effective pressure transmitted, after allowing for the inertia of the reciprocating parts, is the intercept between *de* and *abc*. Hence, for the position 2 of the crank pin, $p = km$ is the force producing a tangential effort at the crank pin. Consequently the tangential effort on the crank pin at 2 (see § 41) is $t = p \sin \theta = p \dfrac{h^2}{c_2}$. This is easily obtained graphically. Take $cf = km$, and draw fg vertical, then $t = fg$.

To draw a diagram of crank-pin effort therefore proceed thus. Divide the crank-pin circle into any number of equal parts. Take A' B' (fig. 107) equal to the length of the semicircle, that is, to πR, and divide it into an equal number of parts. For each of the points 1, 2, 3, . . . determine the value of t as above and set it up at 1' 2' 3' . . . The curve A' n B' through the points so determined is the curve of tangential crank-pin effort. Obviously the area of this curve is equal to that of the original indicator diagram, but the

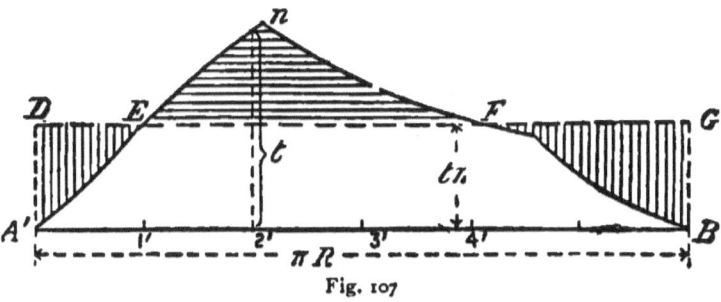

Fig. 107

variation of crank-pin effort during the semi-revolution is quite different from the variation of piston effort. Let t_m be the mean ordinate of this curve determined in the ordinary way. Take A' D $= t_m$, and complete the rectangle A' D G B'. Since the resistance overcome is assumed constant, it must be equal to t_m. Consequently the vertically shaded parts of the diagram correspond to parts of the revolution during which the resistance exceeds the effort and the horizontally shaded part to a portion during which the effort exceeds the resistance. At E the velocity of the engine will be a minimum and at F a maximum. The horizontally shaded area representing an excess of energy must be equal to the vertically shaded areas which represent a deficiency of energy. If E = area of effort curve for one stroke $A'n B'$, or of the rectangle $A' D E B'$ and ΔE = area $E n F$, then the ratio $k = \Delta E/E$ may be called the coefficient of fluctuation of energy, and it is the object of the construction above to determine this.

If there are two engines at right angles, two effort curves must be drawn (fig. 108), the beginning of one effort curve corresponding to the middle of the semi-revolution of the other. Adding the ordinates of the two effort curves, a curve of resultant effort is obtained. The mean ordinate A C of this must be equal to the mean or constant resistance. There will now be two minimum and two maximum velocities in a semi-revolution. The coefficient of fluctuation of energy is now the ratio of one shaded area, such as E G H

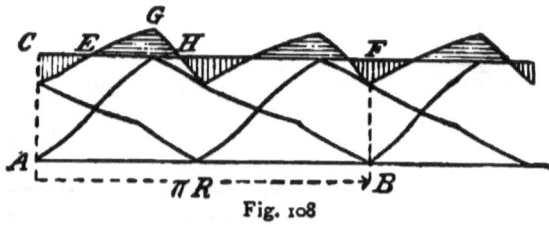

Fig. 108

to the area of the rectangle A C F B. In compound engines with cylinders of different sizes the effective pressure in the H P cylinder must be reduced to the equivalent pressure on the low-pressure piston before proceeding to construct the effort curves, which are to be combined.

111. *Periodical excess or deficiency of energy in an engine driving a pump or air-compressor direct.*—In many pumping engines and air-compressors, the pump-plunger or air-compressor piston is on the same piston rod as the steam piston. Let A B C D be the diagram of steam piston effort corrected, if necessary, for the inertia of the reciprocating parts. Let A C E F be the corresponding diagram of effective resistance at the plunger or compressor piston. If the steam diagram is drawn for pressures in pounds per sq. in.

of the steam piston, the resisting pressures must be reduced to equivalent pressures on a piston of the same area. Then the shaded areas represent the work alternately stored and restored:

$$k = \frac{\Delta E}{E} = \frac{ABC}{ABCDF} = \frac{CED}{AEF}.$$

An interesting case arises in the High Duty Worthington direct-acting pumping engine. In this there is no flywheel. Two small oscillating cylinders are placed at right angles to the line of stroke. During the first half of the stroke the plungers of these are driven in and force back a column of water into a reservoir containing air under pressure. During the second half of the stroke the plungers

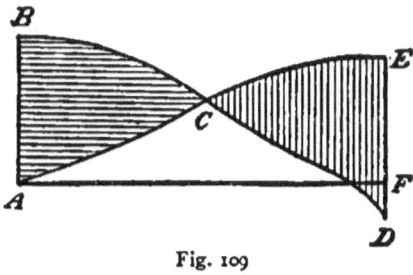

Fig. 109

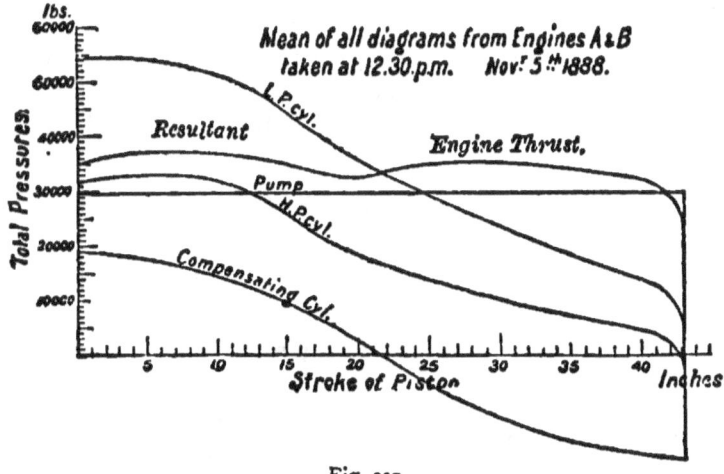

Fig. 110

move out again, giving out just as much work as was previously absorbed. By adjusting the air pressure, an almost exact compensation of the variation of effective steam pressure is obtained.

112. Coefficient of fluctuation of speed.—Let v_1 v_2 be the least and greatest speeds of the flywheel rim at radius R in a period, for instance, the speeds at the points E and F, fig. 107. Then, since $v_2 - v_1$ is small, it is accurate enough to take the mean speed $v = \frac{1}{2}(v_1 + v_2)$. The ratio $(v_2 - v_1)/v$ or approximately $2(v_2 - v_1)/(v_1 + v_2)$ will be called the coefficient of fluctuation of speed, and will be denoted by n. This ratio varies for different kinds of work. The following values correspond with ordinary practice:

		$n =$
Engines doing pumping	.	1/20
,, driving machine tools	.	1/35
,, ,, textile machines	.	1/40
,, ,, spinning machinery	.	1/50 to 1/100
,, ,, electric machinery	.	1/150

113. Weight of flywheel when the coefficients of fluctuation of energy and speed are given.—Suppose an engine is doing E foot pounds of work per stroke and the coefficient of fluctuation of energy, k, for the given conditions of working has been determined. Then $\Delta E = k E$ is known. But from eq. 1 a;

$$\Delta E = \left(W_r + \tfrac{1}{3} W_a\right) \frac{v_2^2 - v_1^2}{2g}$$

$$= \left(W_r + \tfrac{1}{3} W_a\right) \frac{(v_2 - v_1) v}{g}$$

$$= \left(W_r + \tfrac{1}{3} W_a\right) \frac{n v^2}{g}$$

$$W_r + \tfrac{1}{3} W_a = \frac{g \Delta E}{n v^2} = \frac{g k E}{n v^2}. \quad . \quad (2)$$

An equation from which the necessary weight of the flywheel can be determined.

If a is the area of the rim in sq. ft.

$W_r = 2 \pi R a \times 450$ lbs. nearly, from which a provisional estimate of the section of the rim can be made.

CONSTRUCTION OF FLYWHEELS

114. *General construction of flywheels.*—Flywheels up to 10 feet in diameter are sometimes cast solid in one casting. But sometimes the nave is split to relieve the stresses due to contraction in casting. Then rings shrunk on or bolts must be used to connect the two parts of the nave. For flywheels from 10 to 15 feet in diameter a safer construction is to cast the wheel in two halves and connect the portions of the rim by dowels and cotters, and the parts of the nave by bolts or rings shrunk on. Larger wheels are cast with nave, arms, and segments separate. The segments of the rim are connected in various ways by flanges and bolts, dowels and cotters, or by straps and bolts. The arms are sometimes halved on to the rim and bolted, but this cuts away half the section of the rim. More commonly the arms are attached by flanges and bolts inside the rim. The arms are attached to the nave in wedge-shaped recesses in which they are bolted, or, better, the end of the arm is turned taper and fits into a taper hole bored in the nave.

Fig. 111 shows one of the best constructions for built-up flywheels. The nave in this case is keyed on the shaft with four keys, two driven one way and two the other. There are flats on the shaft and key ways in the nave. With four keys some adjustment of the wheel, to centre it, is possible, but it is also usual to bore the nave and fix it by two keys in positions at right angles. The larger ends of the arms are turned and fit in bored recesses in the nave, to which they are cottered. The small end of the arms has a flange to receive the segments, the joints in the segments being over the arms. The segments are connected partly by bolting to the arms, partly by a dowel and cotters. The outer surface of the segments is turned to receive toothed segments, the joints of which come half way between the joints of the rim segments. Sometimes the teeth are cast on the rim segments, but this is not a good plan. It is very

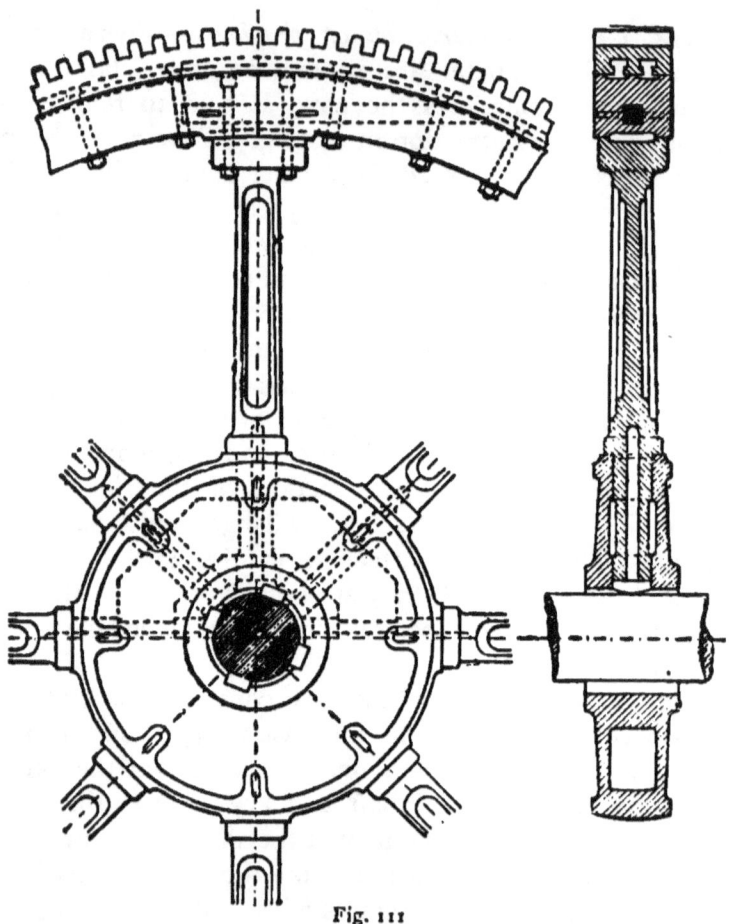

Fig. 111

difficult in that case to secure even pitch of the teeth, and the teeth are liable to be weak and spongy at the roots from contraction of the large mass of metal in the rim in casting. The spur segments in this case have two recesses cast round their inner surface to receive T-headed bolts, by which they are fixed to the rim segments.

115. *Nave with rings shrunk on.*—Fig. 112 shows a flywheel nave split into three parts and connected by two wrought-iron rings shrunk on. The stress in the rings due to shrinking on and to keying is almost indeterminable; consequently a somewhat rough calculation of the stress on the rings is the only one possible, and this matters less, as it is easy to give the rings ample section. Let w be the total weight of the flywheel in lbs., R the radius in ft. to the centre of rim, ω the angular velocity per second, and N the revolutions per minute. The centre of gravity of half the flywheel will be very nearly at the distance $2R/\pi$ from the centre of the wheel. Supposing each half of the wheel revolving at its centre of gravity, and neglecting the connection of the parts of the rim, there will be two radial centrifugal forces—

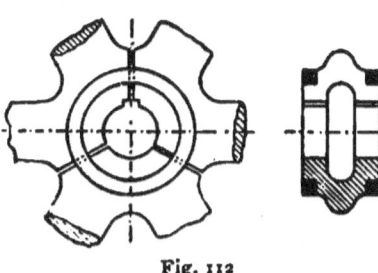

Fig. 112

$$\frac{W}{2g} \cdot \frac{2R}{\pi} \omega^2 = \frac{W}{2g} \cdot \frac{2R}{\pi} \left(\frac{2\pi N}{60}\right)^2$$

$$= \frac{W}{g} R \frac{\pi N^2}{900} \text{ lbs.,}$$

tending to tear the two rings apart at a diametral section. Let a be the section of a ring, and f the working safe stress at normal speed. Then, since there are four sections of the two rings to carry the tearing force,

$$4af = \frac{W}{g} R \frac{\pi N^2}{900}$$

$$a = \frac{WR}{gf} \cdot \frac{\pi N^2}{3,600}.$$

It appears that in actual cases in practice, even on the

extreme supposition that the rings carry all the tension due to the centrifugal force of half the flywheel, f does not exceed 300 lbs. per sq. in., and is often only half this.

Where there are bolts instead of rings shrunk on, the stress on the net section of the bolts at the bottom of the thread is sometimes 2,500 lbs. per sq. in.

116. *Fastenings of rim segments.*—Fig. 113 shows various methods of connecting the rim segments. At A and

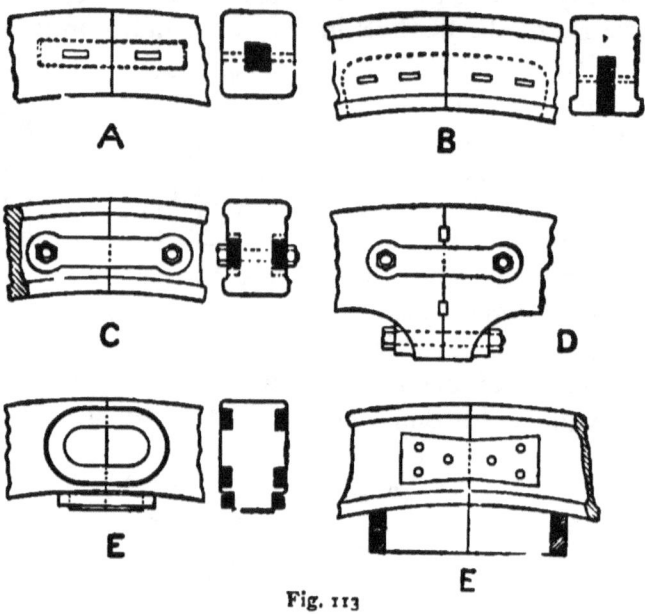

Fig. 113

B wrought-iron dowels are used cottered into the rim. Taking wrought iron to be four times as strong as cast iron, the section of the dowel through the cotter hole should be one fourth of the area of the rim through one of the cotter holes. At C a strap bolted to the rim on each side is used, and at D a similar pair of straps and also a bolt. At E rings shrunk on projections on each side and a third ring shrunk on projections inside the rim are used. The only difficulty with this fixing is that the stress in the wrought-iron straps is not determinable with any accuracy. At E two dovetailed straps riveted to the rim are used, and a ring shrunk on projections inside the rim.

Fig. 114 shows a joint directly over an arm, the joint being made by bolts and flanges. Fig. 115 shows a portion of the rim of a very wide flywheel carrying rope grooves. The arm is attached by T-headed bolts, and other bolts in flanges connect the rim segments.

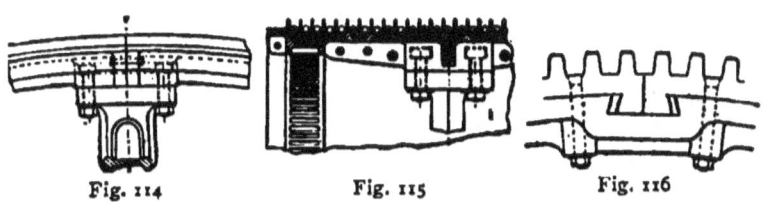

Fig. 114   Fig. 115   Fig. 116

Fig. 116 shows a method of attaching spur segments to the rim segments different from that in fig. 111.

Strength of Flywheels

117. *Approximate calculation.*—Flywheels have, no doubt, very commonly been designed on a very rough theory of the straining actions, and to balance the imperfection of the theory very low working stresses have been assumed. That this rough theory so modified does not always secure safety the accidents which occur sufficiently prove. Except where exceedingly low working stresses are allowed, it can hardly be assumed on grounds of experience that safety is assured. But the adoption of very low working stresses involves making wheels less efficient as stores of energy than they might be if the theory were more perfect and trustworthy. In this as in so many other cases in machinery, safety is often imperfectly secured by unnecessary waste of material, which with more intelligence might be economised without any greater danger, or rather with an increase of safety.

However, it may be useful to give the rough theory first, because any exact theory is complicated, and then to examine more carefully the real straining actions. In the rough theory of flywheels it is assumed that the rim is subjected to a tension due to its centrifugal force, and the arms to a bending moment due to the greatest acceleration or retardation of the rim likely to occur. The bending of the rim and the mutual action of the arms and rim are neglected.

Let R be the radius to the centre of the rim in feet.
,, N be the number of revolutions per min.
,, ω the angular velocity per second $= \pi N/30$.
,, v the linear velocity at radius R in ft. per sec. $= \omega R$.
,, a be the section of the rim in sq. ins.
,, $w = 3\cdot125$ lbs. $=$ the weight of a square bar of one inch section one foot long.

The total weight of the rim is $W_r = 2\pi R a w = 19\cdot63$ R a lbs.

The radial centrifugal force acting uniformly round the rim like a fluid pressure is for each sq. in. of section

$$c = \frac{w}{g}\frac{v^2}{R} = \frac{w}{g}\omega^2 R$$

per foot of arc. The resultant centrifugal force of a semicircle of the rim is per inch of section $2 c R$. Hence the tension in the rim is

$$f = c R = \frac{w}{g} v^2 = \frac{w}{g} \omega^2 R^2 \text{ lbs. per sq. in.}$$

In a flywheel rim of large section the stress will be distributed like that in a thick tube subjected to internal fluid pressure, and therefore will be a little greater than f at the inside and a little less at the outside. Neglecting this, f is the stress in the rim due to centrifugal force, and it depends only on the radius and angular velocity. For cast iron $f = 0\cdot0971 v^2$.

Stress in flywheel rim due to centrifugal force :

$v =$ 30 40 50 75 100 ft. per sec.
$f =$ 87·4 155·3 242·8 546·3 971·0 lbs. per sq. in.

Flywheels

The bending stresses in the rim, due to the centrifugal force of the portions between two arms, may very probably increase these stresses 50 per cent. Some wheels have teeth or toothed segments added to the rim, which increase the centrifugal force without adding to the section which resists it. In some wheels portions of the rim are cut away for cotter or bolt holes. But, allowing for all these causes of weakness, the stresses found in actual wheels are low compared with the safe stress allowed in other parts of machines.

The arms have still to be considered. The arms are principally strained by the bending moment due to variations of velocity of the wheel. Let T be the greatest twisting moment transmitted from the flywheel shaft to the flywheel. In starting, and in cases where steam is suddenly cut off, and the motion is continued by the energy of the flywheel alone, T may have a value equal to the mean twisting moment transmitted in ordinary driving (Part I., § 22, p. 36). Let n be the number of arms, z the modulus of the arm section (Part I. Table V.). Then

$$T = n f_a z$$

where f_a is the stress due to bending in the arm, greatest near the nave. At the rim the section is often only two-thirds that at the nave. It is commonly stated that f_a should not exceed 1,000 to 1,400 lbs. per sq. in. for cast-iron arms. But with the uncertainty as to the magnitude of the moment transmitted through the arms, the calculation is too crude to be very useful.

If d is the diameter of the flywheel shaft at the bearings where the straining action is chiefly a twisting moment, and f the stress in the shaft, then the flywheel arms will be as strong as the shaft if

$$0.2 f d^3 = n f_a z ;$$

if the shaft is of wrought iron and the arms of cast iron, $f = 4 f_a$. Then

$$z = \frac{4}{5} \frac{d^3}{n}.$$

The dimensions of the arm can be determined from the section modulus z (See Table V. Part I.).

118. *Stresses in a homogeneous flywheel running at uniform speed.*—By restricting the case to that of a homogeneous flywheel, it is meant to exclude the case in which the rim and arms are of different materials, and also to exclude the complications arising out of joints in the rim, of strength or stiffness less than that of the rim.

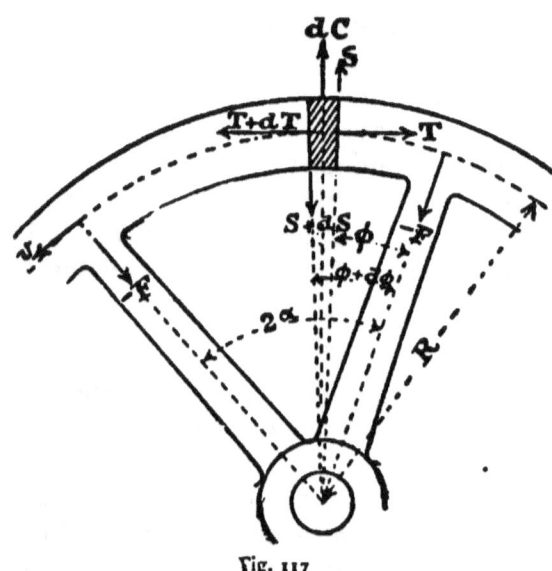

Fig. 117

Let fig. 117 represent a portion of such a flywheel in uniform rotation. Under the action of the centrifugal force there will be a tension in the rim, which, considered alone, would cause an increase of radius uniform all round. This uniform expansion of the rim is, however, hindered at the junctions of the arms. The arms themselves lengthen partly under the action of their own centrifugal force, partly from the pull exerted on them by the rim; but between the arms the rim bends outwards more than at the arms. The result is, the rim takes a form convex to the centre of the wheel at the arms and concave between. The action is similar between each pair of arms.

Let 2α be the angle between the centre lines of two arms. This angle, from conditions of symmetry, remains constant when the wheel is deformed by the stresses acting on it. Let R be the radius to the circle through the centres of figure of the radial sections of the rim, R and all the other dimensions being in feet for simplicity. Let the angular velocity of the wheel be ω, so that the linear velocity at the radius R is $v = \omega R$. Let A be the section (in sq. ft.) of the rim; A_1, the mean section of an arm. Let G be the weight of the material in lbs. per cub. ft., so that G A is the weight of the rim in lbs. per foot of length.

Consider a slice of the rim between two radial sections inclined at an angle $d\phi$. The centrifugal force of this slice is

$$dC = \frac{G}{g} A R d\phi \frac{v^2}{R}$$
$$= \frac{G}{g} A v^2 d\phi \quad . \quad . \quad . \quad (3)$$

At the end faces of the slice act the radial shearing forces S and $S + dS$ and the normal tensions T and $T + dT$. Further, we must suppose at these faces couples of forces producing bending moments M and $M + dM$.

For equilibrium we must have

$$\frac{G}{g} A v^2 d\phi - T d\phi - dS = 0 \quad . \quad . \quad (4)$$
$$dT - S d\phi = 0 \quad . \quad . \quad (5)$$
$$dM - R dT = 0 \quad . \quad . \quad (6)$$

The last equation is obtained by taking moments about the centre of the wheel. From (5) and (6) we get

$$dM - R S d\phi = 0 \quad . \quad . \quad (7)$$

Differentiating (4) with respect to ϕ, we get

$$\frac{d_2 s}{d\varphi^2} = -\frac{dT}{d\varphi},$$

or with the relation in (5)

$$\frac{d_2 s}{d\varphi^2} + s = 0,$$

whence by integration

$$\frac{ds}{d\varphi} = \pm \sqrt{c - s^2}$$

$$\sin^{-1} \frac{s}{\sqrt{c}} = c_1 \pm \varphi$$

$$s = \sin(c_1 \pm \varphi)\sqrt{c}.$$

Consequently from a known relation

$$s = a \sin \varphi + b \cos \varphi \qquad \ldots \qquad (8)$$

where a and b are new constants requiring to be determined. Using this value in (4)

$$T = \frac{G}{g} A v^2 - a \cos \varphi + b \sin \varphi \qquad \ldots \qquad (9)$$

and from (6)

$$M = \frac{G}{g} A R v^2 - a R \cos \varphi + b R \sin \varphi + c. \qquad (10)$$

where c is a new constant.

To determine these constants a, b and c, consider the values of s, T and M for $\varphi = 0$, and $\varphi = 2a$,

For $\varphi = 0$, $s_o = b$.

For $\varphi = 2a$, $s_a = a \sin 2a + b \cos 2a$, and since $s_o = -s_a$, because at all the arms s has the same value,

$$s_o - s_a = 2s_o = -2s = -F$$

if by F we understand the tension in the arm due to the centrifugal force. Hence

$$b = -\frac{F}{2}. \qquad \ldots \qquad (11)$$

$$\frac{F}{2} = a \sin 2\alpha - \frac{F}{2} \cos 2\alpha$$

$$a = \frac{F}{2} \frac{1 + \cos 2\alpha}{\sin 2\alpha} = F/2 \tan \alpha \quad . \quad . \quad (12)$$

Introducing these values of the constants

$$S = -\frac{F}{2} \frac{\sin(\alpha - \phi)}{\sin \alpha} \quad . \quad . \quad . \quad (13)$$

$$T = \frac{G}{g} A v^2 - \frac{F}{2} \frac{\cos(\alpha - \phi)}{\sin \alpha} \quad . \quad . \quad (14)$$

$$M = \frac{G}{g} A R v^2 - \frac{F R}{2} \frac{\cos(\alpha - \phi)}{\sin \alpha} + c \quad . \quad (15)$$

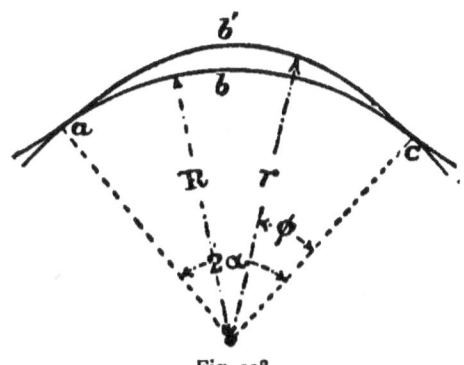

Fig. 118

It still remains to determine the constant c. Under the action of the bending moments the centre line of the rim, originally circular, as at abc, takes a form $ab'c$, fig. 118. The angle α must remain unchanged from conditions of symmetry. Using the ordinary equation to the elastic line, and remembering that r and R differ only by a very small quantity,

$$M = \pm E I \frac{d_2 r}{(R \, d \, \psi)^2} \quad . \quad . \quad . \quad (16)$$

where E is the coefficient of elasticity of the material and I the moment of inertia of the rim section about an axis through its centre of figure. Using the value of M in (15) and integrating

$$\pm \frac{dr}{R\,d\varphi} EI = \frac{G}{g} A v^2 R^2 \varphi + \frac{F}{2} R^2 \frac{\sin(a-\varphi)}{\sin a} + c R \varphi + \text{const.}$$

But for $\varphi = 0$

$$\frac{dr}{R\,d\varphi} = 0$$

$$\text{const} = -\frac{F}{2} R^2.$$

Also for $\varphi = 2a$, $\dfrac{dr}{R\,d\varphi} = 0$, and hence, inserting the value just found and reducing

$$c = -\frac{G}{g} A R v^2 + \frac{F R}{2a}. \qquad . \qquad . \qquad (17)$$

Putting this value in (15) we get finally

$$M = \frac{F R}{2a} - \frac{F R}{2} \frac{\cos(a-\varphi)}{\sin a} \qquad . \qquad . \qquad (18)$$

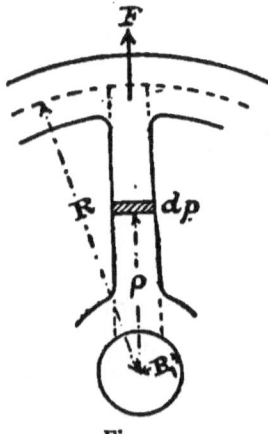

Fig. 119

Equations (13), (14) and (18) now give the shearing force S, the tension T and the bending moment M at any section of the rim, provided that the value of F is determined. To find this the stretching of the arm must be determined. Consider a slice of the arm between sections at radii ρ and $\rho + d\rho$, fig. 119.

Let A_1 be the mean section of the arm. Then the weight of the slice is $G A_1 d\rho$; its velocity is $v\rho/R$; and its centrifugal force is

$$d c_1 = -\frac{G}{g} A_1 d\rho \frac{v^2 \rho}{R^2} \qquad . \qquad . \qquad (19)$$

The total stress on the section is therefore

$$C_1 + F = -\int_R^\rho \frac{G}{g} A_1 \frac{v^2 \rho}{R^2} d\rho + F$$

$$= \frac{G}{g} A_1 \frac{v^2}{R^2} \frac{R^2 - \rho^2}{2} + F.$$

The extension of the arm by this varying tension, which may be denoted by ΔR, is

$$\Delta R = \int_{R_1}^R \frac{G}{g} A_1 \frac{v^2}{R^2} \frac{R^2 - \rho^2}{2} \frac{d\rho}{A_1 E} + \int_{R_1}^R \frac{F\, d\rho}{A_1 E}$$

where for simplicity the arm is supposed to extend from the eye of the nave, to the centre of the rim. Integrating and reducing

$$\Delta R = \frac{G}{gE} \frac{v^2 (R - R_1)}{3} \left[1 - \tfrac{1}{2} \frac{R_1}{R} - \tfrac{1}{2} \left(\frac{R_1}{R}\right)^2 \right]$$

$$+ \frac{F(R - R_1)}{A_1 E} \quad . \quad . \quad . \quad . \quad (20)$$

As R_1 is small compared with R, this may be taken with accuracy enough to be

$$\Delta R = \frac{G}{gE} \frac{v^2 R}{3} + \frac{FR}{A_1 E} \quad . \quad . \quad . \quad (21)$$

But from the extension of the rim by the tension T a second equation can be obtained. The length of rim between two arms is $2 R a$, and putting $\Delta (2 R a)$ for the extension of this,

$$\Delta (2 R a) \quad \int_0^a \frac{T R\, d\phi}{A E}.$$

But $2 a$ remains unchanged, so that $\Delta (2 R a) = 2 a\, \Delta R$. Hence, using the value of T in (12), and integrating

$$\Delta R = \frac{R}{E} \left\{ \frac{G}{g} v^2 - \frac{F}{2 A a} \right\} \quad . \quad . \quad (22)$$

By elimination in equations (22) and (20) we get

$$F = \frac{2}{3} \frac{G}{g} v^2 \bigg/ \left(\frac{1}{A_1} + \frac{1}{2 A a} \right)$$

$$= \frac{2}{3} \frac{G}{g} v^2 \frac{2 A_1 A a}{A_1 + 2 A a} \quad . \quad . \quad . \quad (23)$$

CHAPTER IX

VALVES, COCKS, AND SLIDING VALVES

Valves

119. In all machinery put in motion by the action of a fluid (water or steam) or employed in pumping fluids, valves are required to regulate the admission and discharge of the fluid. With reference to the mode in which the motion of valves is obtained, they may be divided into four classes : (1) Valves opened and closed by hand ; (2) Valves opened and closed by independent mechanism ; (3) Valves opened and closed by mechanism so connected with the machine as to render the times of opening and closing synchronous with the motions of the machine ; (4) Valves opened and closed by the action of the fluid.

In the case of valves opened and closed by the action of the fluid, the valve opens if the fluid moves in one direction and closes if it moves in the other direction. Independently closed valves, or valves controlled by mechanism, are more generally arranged to close a passage to the fluid, in whichever direction it tends to move.

Another convenient classification depends on the way in which the valve moves relatively to its seat. Thus we have : (1) flap or butterfly valves, which rotate in opening ; (2) lift valves, or puppet valves, which rise perpendicularly to the seat ; (3) sliding valves, which open by moving parallel to the seat.

AUTOMATIC VALVES

120. The conditions to be fulfilled by a good automatic valve are these : (1) accurately timed opening and closing ; (2) small resistance to the flow of the fluid ; (3) tight closing. The tightness of the valve is mainly a question of fitting and the selection of suitable material for the valve seat. To secure small resistance to flow the valve must open fully, and this is a requirement somewhat in conflict with the condition of closing accurately at the proper time. The importance of the resistance at the valve varies very much in different cases, and a good deal of resistance may be allowed in some cases rather than incurring the risk of delayed closing of the valve, and consequent shocks. The question of the closing of the valve will be dealt with later. Very often, in quick-acting pumps, springs are added to hasten the closing of the valve at the change of stroke of the pump.

Fig. 120 gives a general view of the types of automatic valves. At B is the simplest form of leather-flap valve, the leather being stiffened with metal plates. Such valves work very well under low pressures, and when the number of beats per minute is not too great. A similar valve of brass is shown at C hinged to a metal seating a, which is fixed by wood wedges b, and can be lifted out through the aperture covered by the plate c. This is a common form of condenser foot valve. At A is an india-rubber valve which acts in a similar way. The india-rubber valve is in the form of a circular disc covering a gridiron seating. The bars of the gridiron support the flexible india-rubber, and should be placed closer the greater the pressure on the valve. The valve is held by a bolt at the centre, and takes a dished form in rising. A guard plate prevents the valve from rising too high. At D are two flap valves placed back to back, and then termed 'butterfly valves.' They beat on a brass seating, and have a guard to limit the lift. G is a mushroom lift

194   *Machine Design*

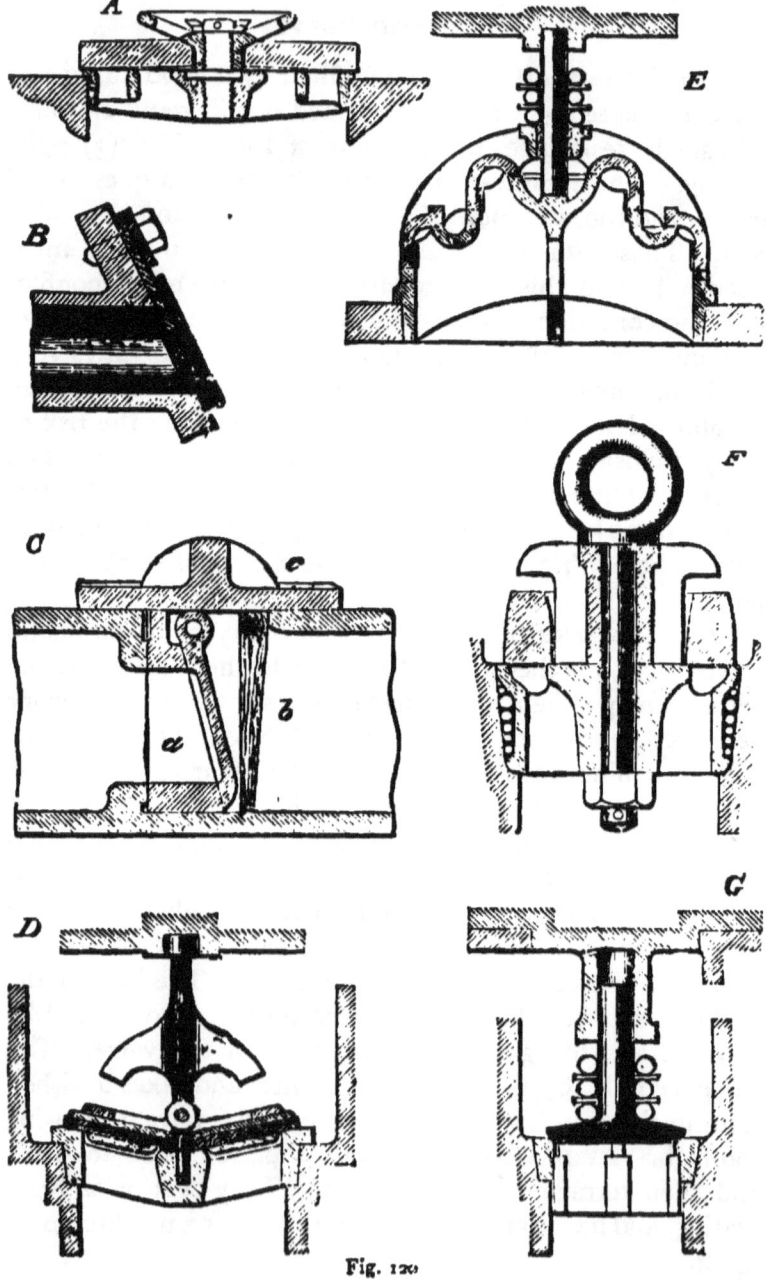

Fig. 120

or puppet valve; it consists of a metal disc fitting accurately a flat or, more commonly, a conical seat. It is guided, in rising, either by a spindle or by three feathers underneath the valve, sliding in the valve seat. The bearing surface of the seat must be made narrow (about $\frac{1}{16}$ to $\frac{3}{16}$ inch), or it will not be tight. Consequently, as the valve is made larger, the intensity of the pressure on the seat increases. Hence there is a difficulty in using valves of this kind of large size. Further, the lift of such a valve should be proportional to its diameter, to give adequate waterway when open. But when the valve is large and the lift great, the shock of the valve in closing becomes severe and damages the valve and seating faces. At F is a ring valve, in which these objections are to some extent obviated. When open the water escapes at both the inside and outside edges of the valve, and hence for a given waterway only half as much lift is necessary as would be required by a mushroom valve of the same diameter. At E is a valve used for large pumping engines, which consists of two ring valves, and gives four edges for the escape of the water. At G and E the valves are provided with an arrangement to accelerate the closing of the valve, introduced by Mr. H. Davey. This consists of india-rubber washers, forming a spring. The valve has a certain amount of free lift, and the rest is obtained by compression of the india-rubber. In G the spindle often has 3 flats planed, so as to form passages for the fluid when the valve lifts.

Fig. 121

121. *Flap or butterfly valves.*—Fig. 121 shows the simplest form of flap valve, formed of a leather disc, strengthened and stiffened by two plates of iron, of brass or of lead, which at the same time give weight enough to the valve to close rapidly, when the

pressure beneath it ceases. A butterfly valve consists of two flap valves placed hinge to hinge, or sometimes edge to edge. In the latter position the direction of motion of the fluid is less interfered with. The lifting of the valve is usually restricted to an angle of 30° or 40°; width of seat $\frac{1}{5}\sqrt{D}+\frac{1}{8}$ or in some cases 0·1 D. Valves of this kind are most commonly lifted by the fluid.

122. *India-rubber disc valve.*—A form of valve very extensively used for condensers and pumps consists of a

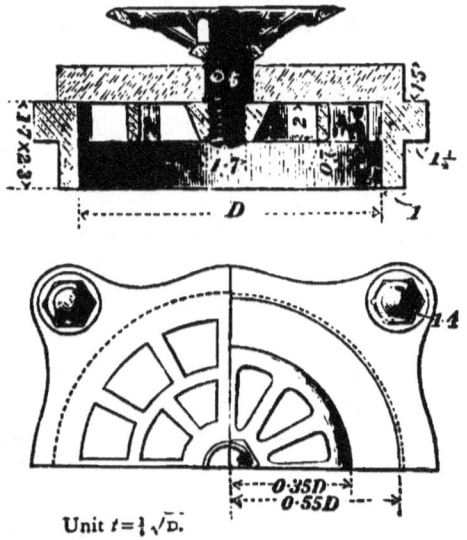

Fig. 122

circular disc of india-rubber, secured by a bolt at the centre, and resting on a brass grid which forms the seating. The india rubber, being flexible, lifts easily from the grating, when any fluid pressure is applied beneath it, and closes again readily, and without violent shock, when the reflux begins. To prevent the india-rubber rising too high, a perforated guard plate is placed over the valve. Figs. 122 and 123 show two of these valves. In one the valve seat is attached to the cast-iron casing of the condenser by bolts, and the india-rubber and guard plate are attached to it by a stud. In the other the seating, india-rubber and guard plate are all secured by the same central bolt, which bears against a cross bar on the other side of the casing to that on which the valve is placed. In each of these figures the valve and guard plate are removed from one half of the plan, in order

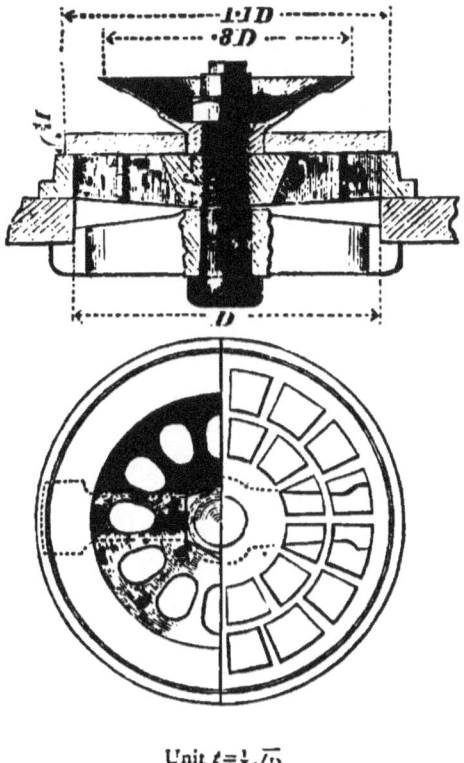

Unit $t = \frac{1}{8}\sqrt{D}$
Fig. 123

to show the grating on which the valve rests. The india-rubber should not be too thin, $\frac{3}{4}$ inch to $\frac{7}{8}$ inch thickness is sufficient. The apertures of the grating should be so small that there is no great flexure of the india-rubber, and the grating area should be so large that the pressure with valve closed does not exceed 40 lbs. per sq. in. It is more satisfactory to use several valves of 7 inches to 9 inches diameter than to use a single large one. India-rubber valves should not be used for pressures above 100 lbs. per sq. in.

The guard plate is often of spherical form of a radius equal to the diameter of the valve. In other cases the guard plate is flat, and the india-rubber disc lifts bodily on the central stud. It is well to give a small lift in all cases before the valve reaches the guard plate. It is now common to place a spiral spring above the valve to insure more rapid closing.

The throttle valve used on many engines, which consists of a circular or square metal disc, capable of turning about a shaft passing through it in the direction of a diameter, is a kind of double-flap valve. The disc is placed in a pipe, and closes the passage-way when placed across the pipe, whilst it offers little resistance when parallel to the axis of the pipe. This valve is an imperfect equilibrium valve, the pressure on one half partly balancing the pressure on the other, so that the force required to move the valve is only equal to the difference of these two pressures. The equilibrium is exact, however, only while the valve is shut or so long as there is no sensible current passing it. If a rapid current is established, the pressure on that half of the valve which first deflects the current is greater than on the other half, thus tending to close the valve.

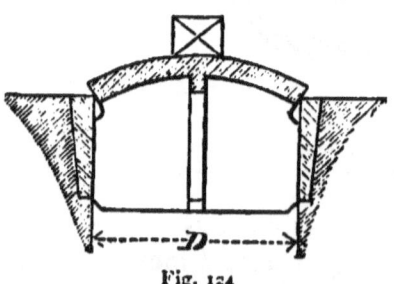

Fig. 124

123. *Lift or puppet valves.* — These are very various in form, the simplest being a circular disc, usually of metal, with a flat or bevelled edge, which fits a circular metal seating (fig. 124). These valves are generally placed with the axis of the valve vertical, so that their weight tends to keep them closed; but

they may be otherwise placed if springs or rods are used to close them.

Let D be the diameter of the valve seating, then the waterway through the valve seating is $\frac{\pi}{4} D^2$. If h is the height of lift, the waterway round the edge of the valve when open is $\pi D h$. In order that these two may be equal, the lift h must be one-fourth of the diameter D. It is only possible to give so great a lift in small valves, because with a great lift the valve acquires too much velocity in closing, and there is a violent shock, causing vibration and damage to the valve seat. It is necessary often to restrict the lift to a less amount than would otherwise be desirable, and then the resistance to the passage of the fluid is increased. Hence it is better to use two valves of diameter 0·707 D than a simple valve of diameter D. For pumps the mean velocity of water through the seating is generally limited to 3 feet per second. Let s be the width of the bearing surface of the valve and seat, measured on a plane at right angles to the direction of lift of the valve. Then $\pi D s$ is nearly the effective area of the bearing surface. If p is the greatest difference of pressure on the two sides of the valve, then $p/\pi D s$ is the crushing pressure on the narrow valve seating. Experience shows that it is not desirable that this pressure should exceed certain limits.

Greatest pressure on surface of valve seat

Gun-metal	2,000 lbs. per sq. in.
Phosphor bronze	3,000 ,, ,, ,,
Cast iron	1,000 ,, ,, ,,
Leather and india-rubber	700 ,, ,, ,,

These values determine the minimum width of valve seat. In valves for water the part of the seat on which the valve beats is sometimes of lignum vitæ, sometimes of leather, sometimes of an alloy of lead and tin. For air-compressor valves tin is used, for steam gun-metal.

Fig. 125 shows a conical disc valve and casing. The valve is guided in rising and falling by three feathers which fit the cylindrical part of the seating, and are shown in the

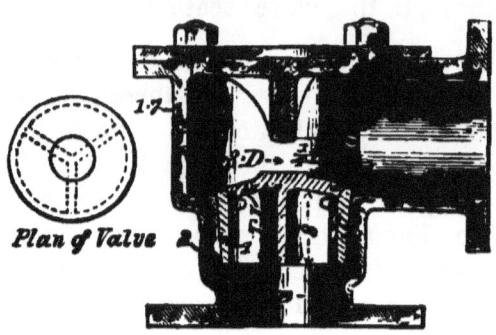

Unit $t = \frac{1}{3}\sqrt{D}$
Fig. 125

plan of the valve. The lift of the valve is limited by a projection on the cover of the casing. The fitting part, or face of the valve, should be narrow, as it is then more easy to make it tight. It must, however, present area enough to

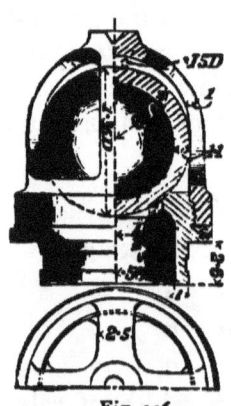

Fig. 126

resist deformation by the hammering action of the valve. The inclination of the face of the valve is usually 45° with the axis of the valve. The horizontal projection of the bevil may have a width $\frac{1}{6}\sqrt{D} + \frac{1}{16}$, or may be determined for the pressure. Conical disc valves may either be actuated by the fluid pressure or by hand. In the latter case they are opened and closed by a screwed rod.

Fig. 126 shows a ball valve, which acts in precisely the same way as a disc valve, except that, as the surface of the ball is accurately spherical, it fits the seating in every position. The only guide required is, therefore, an open cage, which limits the play of the valve. Such valves are often used for small fast-running pumps. To lighten the ball it is often made hollow.

The proportions of valves depend partly on the diameter. Thus the area of the waterway must be constant, and the linear dimensions of the casing are proportional to the valve's diameter. But the thicknesses are in most cases excessive as regards strength, especially in small valves, and do not increase in the same proportion as the diameter. For these the empirical proportional unit

$$t = \tfrac{1}{8} \sqrt{D}$$

will be adopted, where D is the diameter of the valve.

MECHANICALLY-CONTROLLED VALVES

124. *Screw-down valve.*—Fig. 127 shows a lift valve arranged to be worked by hand instead of automatically. Such screw-down valves are now largely used in place of cocks, being tighter and free from liability to stick. The valve rod passes through a stuffing-box, and is made so long that the screwed part does not come in contact with the stuffing-box packing. The valve is fixed on the end of the rod by a pin in such a way that it can turn round. It does not then grind on the seat when screwed down. For steam the valve faces are of gun-metal; for water one of the valve faces may be of leather, of india-rubber, of vulcanised fibre, or, in large valves, of hippopotamus hide. When used for water, it is convenient that the pressure

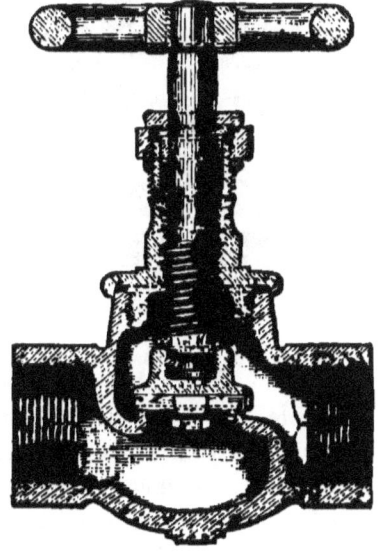

Fig. 127

should be on the under side of the valve when closed; then there is less danger of leakage at the stuffing-box.

125. *Double-beat or Cornish valve*.—The objection to a great lift in metal valves has already been mentioned. In the double-beat valve, two valve faces are obtained in the same valve, and two annular areas are opened when the valve lifts. For a given area of opening, the lift is only about one-half that of a simple lift valve of the same diameter. Fig. 128 shows a Cornish valve for a pumping engine. This valve is raised and lowered by a cam acting on an arrangement of levers. The lower seating is carried directly by the steam-chest. The upper seating is carried by four feathers or radiating plates cast with the lower seating. The valve itself is ring-shaped. Since the two valve faces are nearly of the same diameter, another subsidiary advantage is gained in this form of valve. The valve is pressed down on its seat, partly by its weight, partly by the steam pressure acting on one side of it. If the valve were a simple disc valve, the steam pressure would act on an area $\frac{1}{4} \pi D^2$, where D is the diameter of the valve. As the valve is annular, however, the steam presses only on the area $\frac{1}{4} \pi (D_1^2 - D_2^2)$ where D_1 and D_2 are the diameters of the two faces. Hence the valve is easily lifted against the steam pressure. If the valve is to lift so that the area through the valve is equal to the area through the lower seating, the height of lift must be determined by the equation

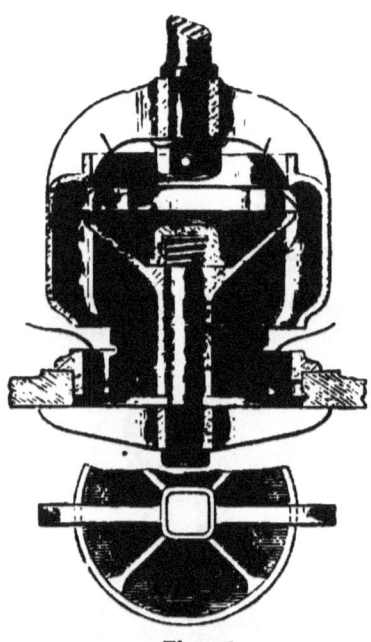

Fig. 128

$$\pi (D_1 + D_2) h = \frac{\pi}{4} D_1{}^2$$

$$h = \tfrac{1}{4} \frac{D_1{}^2}{D_1 + D_2}.$$

125a. *Sir W. Thomson's valves.*—In the case of most valves closed by a screw, the valve comes down on the seating always in the same position, and having once touched the seating any further motion is arrested. If there is any obstruction on the seating it at once becomes jammed between valve and seat, and if the valve and seating are indented, the indents always come together again whenever the valve closes. In some automatic pump valves arrangements are made causing the valve to twist a little in rising, so that it does not always beat on the same place. It occurred to Sir W. Thomson that a valve might be arranged so that there was a definite grinding action of the valve on its seating every time it was closed, this grinding tending to keep the valve and seating true, and to obliterate any accidental damage.

Fig. 128*a* shows a 2-inch check valve arranged in this way, which is manufactured by the Palatine Engineering Company, Liverpool. The stuffing-box is also got rid of, so that there is no perishable material in the valve. The valve is shown at *v*, and is quite loose from the screw plug above. Two lugs on the valve project up into a slot between cheeks, *a*, in the lower part of the screw plug, so that as the screw plug turns the valve must turn with it. A pin *b* with a spiral spring resting on a shoulder presses the valve down perfectly centrally. As the screw plug *s* is turned, the valve is forced down by the spring, and after it reaches its seat it still goes on turning till the central pin *b* jams between the valve and screw plug. During this turning the self-grinding of the valve faces goes on. In the upper part of the valve casing is a conical plug *c* ground accurately into the casing. This replaces the stuffing-box, and is held up by a phosphor

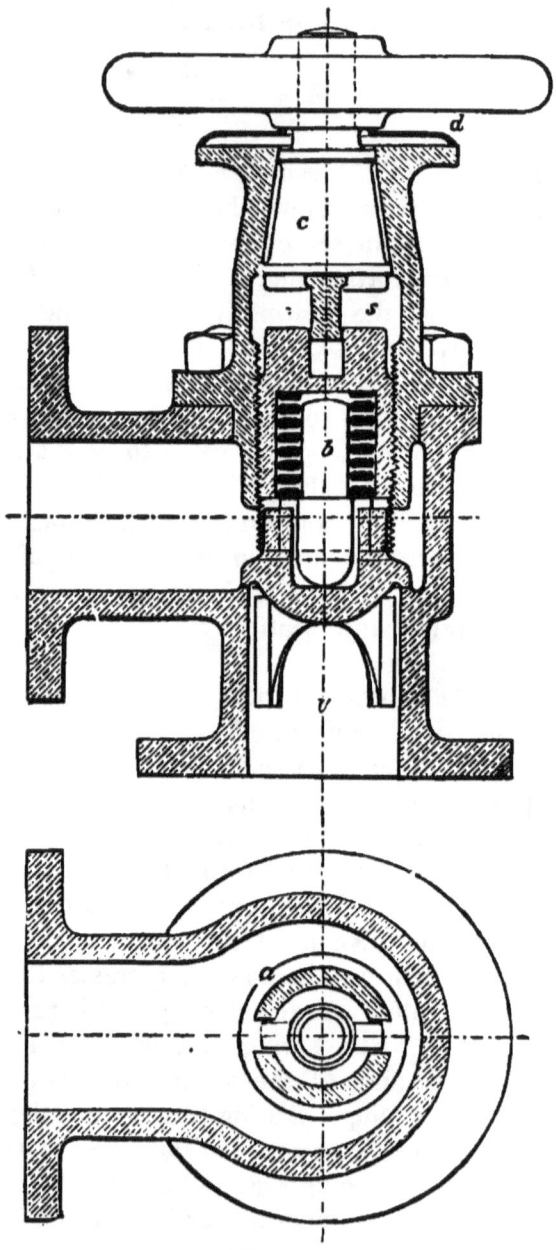

Fig. 128a.

Valves, Cocks, and Sliding Valves

bronze disc spring d. A loose link between the conical and screw plugs transmits the twisting effort from the hand-wheel to the screw plug.

It will be seen that the valve is free from any packing needing to be replaced, the parts are all strong and free from any liability to be bent by rough usage, and they can all be completely fitted by machinery.

The same principle has been applied for ordinary draw-off cocks for houses. One of these is shown in fig. 128b. The valve is separate from the handle and screw plug, but at the same time is forced to rotate with it, having two lugs which slide in slots in the screw plug. A phosphor bronze spiral spring in the cylindrical recess presses down the valve to the bottom of its range. In closing the draw-off cock, the loose valve first comes down on the seat. Continuing to turn the handle, the spiral spring is compressed and at the same time the valve is rotated on its seating while subject to the pressure of the spiral spring. In this way a slight grinding action occurs, tending to keep the valve and seat true, every time the valve is closed. In use these valves are found to keep in perfect order, and similar draw-off cocks have been opened and closed by mechanical means at a rate and for a time which would represent fifty years of ordinary wear. At the end of this trial both valve and seating were in perfect order.

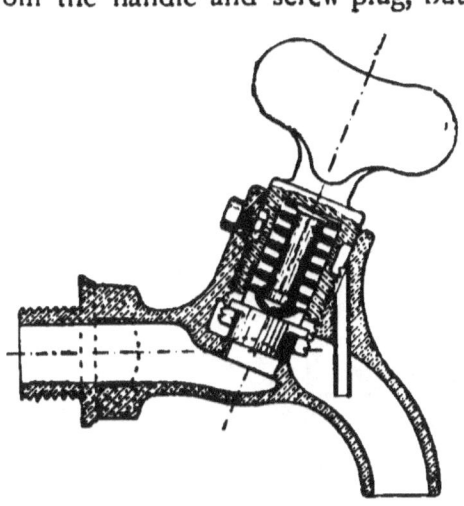

Fig. 128b

Besides the self-grinding action, the stuffing-box is got

rid of in this draw-off cock. A small pipe leads any leakage which passes the screw thread back into the current of water flowing out of the cock. Thus there is no perishable material or packing in the cock, and there is no reason it should not remain permanently in perfect order without any regrinding or attention.

126. *Controlled hydraulically-moved valves.*—In pressure engines and some other cases it is convenient to use valves moved by the action of the fluid, but under conditions which place under control the times at which the valves open and shut. The general principle is to connect the valve to a piston in the valve chamber, and to use a small subsidiary slide valve to admit or release the pressure on the back of the piston.

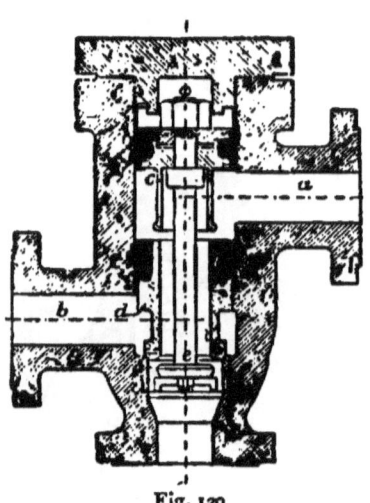

Fig. 129

Fig. 129 shows a valve of this type designed by Mr. Henry Davey ('Proc. Inst. Mech. Eng.' 1880, p. 245). The water is admitted to the pressure engine through the passage a, and is exhausted through the passage b. The valve chamber contains an ordinary puppet valve e, rigidly attached to the piston c, and a second valve d which has two conical seatings, an inner and outer, on its bottom surface, and which itself forms a piston or plunger. The under surface of the piston c and the upper surface of the valve d are always subjected to the water pressure. To the upper surface of the piston c the pressure is alternately admitted and released by a small subsidiary slide valve. In the position shown the pressure is on the top of piston c, and water is being admitted through the centre of valve d to the engine cylinder. If now the pressure on the top of c is released, it will rise, lifting the puppet valve e so as to close the admission of water to the engine cylinder, and afterwards lifting d and e together so as to open the exhaust. On readmitting water to the top of c, the reverse action takes place. As c descends the exhaust first closes, from d coming down on its outer seating, and then e opens. The areas of d are so arranged that there is always an unbalanced pressure acting downwards.

SLIDING VALVES AND COCKS

127. *Sliding valves.*—Sliding valves are more commonly used than any others for stop valves, which are opened and closed by hand; they may be divided into two classes: (1) those with plane faces and seats; (2) those with cylindrical or slightly conical faces and seats. The former class includes engine slide valves and the sluices, often very large, which are used as stop valves on water mains. The latter class includes the hand-worked valves commonly known as 'cocks.'

Sliding sluice valve.—Fig. 130 shows a form of sliding sluice valve often used on water mains and often constructed of very large size. At first these valves were made with only one face, and were kept tight by the water pressure on the back. Now they are almost always double faced, as in fig. 130, and close the pipe against flow in either direction. The double-faced valve is due to Mr. Nasmyth. The valve is wedge-shaped, so as to lift without much friction when once started. It is desirable to make the screw of gun-metal to prevent oxidation. In large valves the screw and its nut are outside the valve casing, where they are more accessible and more easily lubricated.

128. The term 'cock' is sometimes used for any valve opened or closed by hand, but it is more properly restricted to valves which are nearly cylindrical, and which rotate in seatings of the same figure. In ordinary cocks, the seating is a hollow, slightly conical casing, and the valve, which is termed a 'plug,' fits accurately in the seating. The passage-

way for the fluid is formed through the plug. By rotating the plug in one direction its apertures are made to coincide with the entrance and discharge orifices of the casing. The

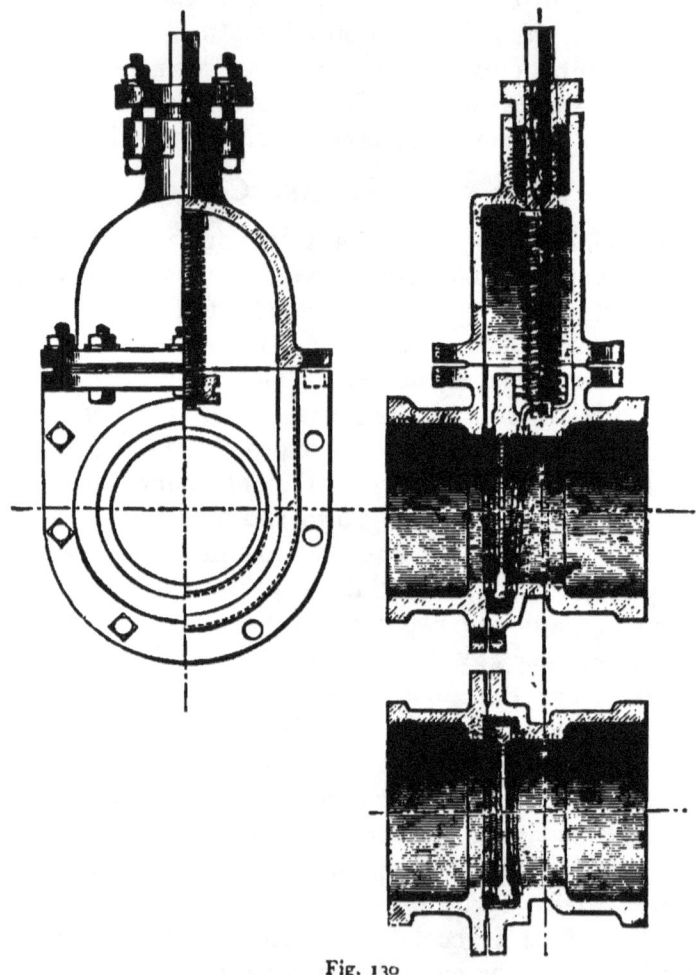

Fig. 130

cock is then open. By rotating it in the other direction the holes in the plug are brought over blank parts of the casing and the cock is closed. The slight taper given to the plug enables it to be accurately fitted, by turning and grinding, to its seating, and it can from time to time be refitted. Each time it is refitted the plug sinks a little lower in the casing. If the plug were cylindrical this refitting would be impossible. The objection to the use of cocks

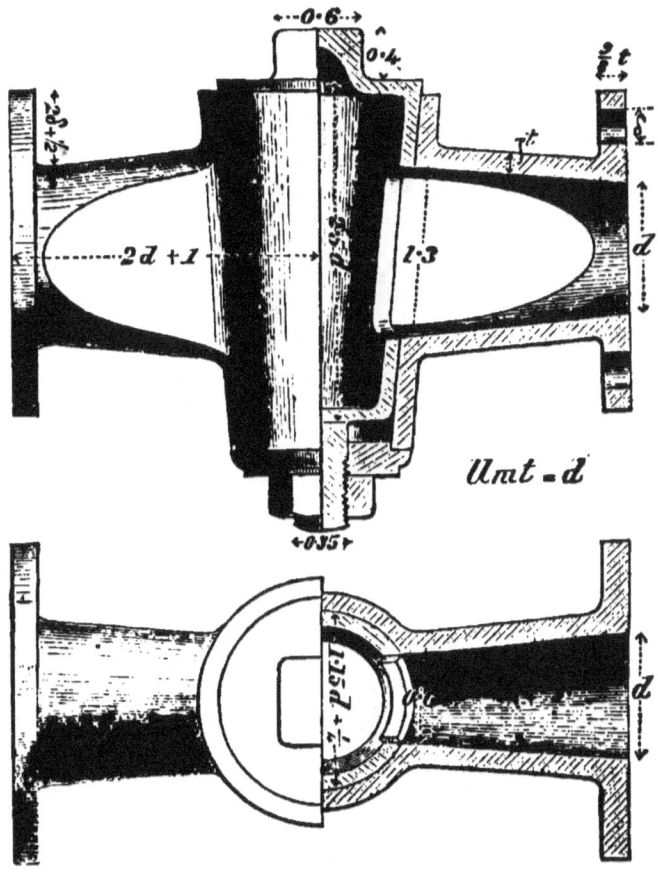

Fig. 131

in many cases, especially for pipes of large size, is that a good deal of power is required to move them, and this is partly due to the conical form, which increases the friction.

The simplest cocks have a solid plug, which is kept in place by a screwed end. When the cock is small, the casing

210   *Machine Design*

has a screwed socket on one side and a screwed end on the other, for the attachment of the cock to the pipes with which it is connected. But in larger cocks, the inflow and outflow orifices are provided with flanges (fig. 131).

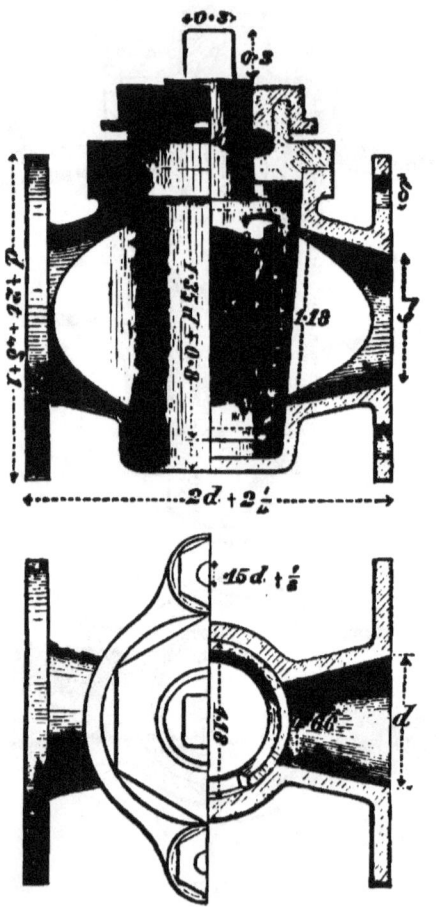

Fig. 132

For small brass cocks, with socket and spigot ends, the following proportions may be adopted :—

Diameter of waterway of cock = d

Diameter of plug at centre $= 1{\cdot}15d + \frac{1}{4}$
Height of hole in plug $= 1{\cdot}3d$
Width of hole in plug $= 0{\cdot}6d$
Total length of tapered part of plug $= 2{\cdot}5d$ to $3d$
Side of square for handle $= 0{\cdot}7d$
Height of square for handle $= 0{\cdot}4d$
Thickness of metal $= 0{\cdot}2d + \frac{1}{16}$
Diameter of plug screw $= 0{\cdot}35d$
„ screwed end $= d + \frac{5}{16}$
Internal diameter of socket end $= d + \frac{3}{16}$
Total length $= 3{\cdot}3d$
Taper of plug $= 1$ in 12 to 1 in 9 on each side.

For cocks with flanged ends, like that shown in fig. 131, the proportions are the same. When the cock is not very small the thickness is best obtained from the rule—

$$t = \tfrac{1}{8}\sqrt{d} + \tfrac{3}{16} \text{ for cast iron}$$
$$= \tfrac{1}{15}\sqrt{d} + \tfrac{3}{16} \text{ for brass.}$$

Some proportions are marked on the figure.

129. Large cocks connected with boilers, and in situations where failure would be dangerous, are best made with closed ends, as shown in fig. 132. The proportions of cocks of this description are a little different.

Diameter of waterway $= d$
Thickness of plug (brass) $= 0{\cdot}12\sqrt{d} + \tfrac{1}{8}$
„ „ (cast iron) $= 0{\cdot}18\sqrt{d} + \tfrac{1}{4}$
„ shell (brass) $= 0{\cdot}18\sqrt{d} + \tfrac{1}{8}$
„ „ (cast iron) $= 0{\cdot}25\sqrt{d} + \tfrac{1}{4}$.

The shell may be reduced to the same thickness as the plug in parts which do not require to be turned.

Diameter of plug at centre $= 1{\cdot}18d$
Size of openings in plug $= 1{\cdot}18d \times 0{\cdot}66d$
Overlap of plug at top and bottom $= 0{\cdot}08d + 0{\cdot}4$

Depth of stuffing-box $=\frac{1}{8}d+\frac{1}{2}$
Depth of gland $=\frac{1}{20}d+\frac{1}{4}$
Diameter of studs in cover, $\frac{1}{8}d+\frac{1}{8}$
Taper of plug $=1$ in 12 on each side.

Some other proportions are marked on the figure.

130. *Dewrance's asbestos packed cocks.*— Ordinary taper plug cocks, when used for steam purposes, often give trouble by sticking fast or leaking. The expansion of the plug causes sticking, and grit between the surfaces causes abrasion and leakage. Hence Messrs. Dewrance have introduced cocks packed with asbestos like a stuffing-box. The plug is smaller than the shell, so that there is room for expansion. The plug is packed with strips of asbestos in recesses running down the shell. There are four grooves down the shell, and a recess also at the bottom under the plug.

THEORY OF THE ACTION OF AUTOMATIC VALVES

131. An automatic valve is one which is opened by the action of the fluid pressure and closed by the action of its own weight or by a spring. To open the valve there must be an excess of fluid pressure below it. While open, the effective closing force, consisting of the weight of the valve (immersed in water) together with any applied force, such as a spring pressure, must be balanced by the forces due to the deviation of the water round the valve, producing an excess of pressure below, and probably, in consequence of the curvature of the stream lines, a diminished pressure at the back. Lastly, as the flow through the valve diminishes, the effective closing pressure should be sufficient to bring it to its seat, before any reflux can take place.

Taking the case of a pump valve, to be more definite, the valve ought to open and close immediately at the turn of the stroke of the pump bucket or plunger. It should cause as little loss of head as possible while open; it should be tight against leakage when closed. The last condition depends chiefly on accurate workmanship. If the velocity through the valve is not excessive (usually 2 to 6 feet per second), the resistance of the valve is not generally a serious item of loss of efficiency. Consequently the accurately-timed opening and closing of the valve are the considerations practically most important.

For the valve to open easily and fully the effective closing force must be as small as possible. On the other hand, if the closing force is too small, the valve closes late, and thence occur shocks, the most serious evil arising in the use of valves. Further, if the valve is of a given weight and lifts during flow to a given height, its closing, due to the acceleration caused by the effective closing force, requires a definite time. Consequently for any valve there will be a limit to the number of strokes per minute of the pump, beyond which the time allowed for closing will be insufficient, the valve will close late, and shocks will arise. According to experiments of Bach, it appears that it is prejudicial to try and get rid of shock by limiting the lift of the valve by fixed stops. Quicker closing can only be secured by increasing the effective closing force.

A remarkable series of observations has been made by Prof. Riedler [1] on the action of pump valves, of very various kinds. He used an indicator somewhat like an engine indicator to obtain curves of pressure during the action of the valves. It is a result of these observations that he has traced shocks, caused by the action of valves in pumping engines, chiefly to the suction valves, and he regards the accurate action of these as of the highest importance. Suppose the suction valve closes late, that is, after the pump plunger has returned through part of its stroke and acquired some velocity; then, when at last the suction valve closes and the delivery valve lifts, the entire

---

[1] 'Indicator-Versuche an Pumpen.' Von A. Riedler. München, 1881.

column of water above the delivery valve has to be suddenly accelerated. The shock is greater, the greater the velocity of the plunger at the moment and the greater the length of the column. Very heavy valves begin to close before the turn of the stroke, and then shocks are less possible. Quietness of action is secured, though at the expense of some additional resistance at the valve.

132. *The excess of pressure due to the area of valve seat.* —If we suppose a valve, fig. 133, to fit its seat so tightly that there is no pressure between the faces of valve and seat, the pressure p_1 above the valve acts on a larger area than the pressure p below the valve. Neglecting the weight of the valve, at the moment when the valve opens,—

$$p\, d^2 = p_1\, d_1^2$$

$$p - p_1 = p_1 \frac{d_1^2 - d^2}{d^2},$$

where $p - p_1$ may be termed the excess pressure due to the area of valve seat. If the assumption that there is no pressure between the

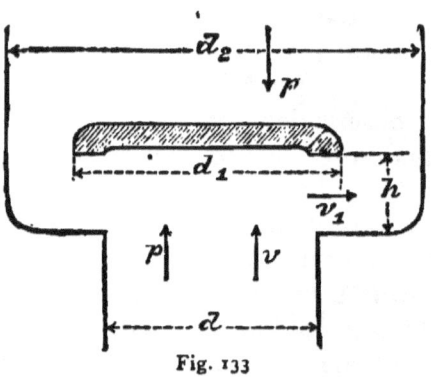

Fig. 133

valve and seat were true, this excess pressure would be in many cases important, and, in fact, it has been supposed to have a great influence on the action of the valve. Experiments with indicators, especially those of Prof. Riedler, show that this excess pressure does not exist, or at all events is much smaller than it would be if the above assumption were true. Suppose, however, there is a fluid layer between the faces of valve and seat. Then the pressure must vary from p at the inner to p_1 at the outer edge, and its mean value cannot be very different

Valves, Cocks, and Sliding Valves

from $\frac{1}{2}(p+p_1)$. Then, still neglecting the weight of the valve, when the valve opens,—

$$p d^2 + \tfrac{1}{2}(p + p_1)(d_1{}^2 - d^2) = p_1 d_1{}^2$$

which is only satisfied if

$$p = p_1,$$

so that there is no excess pressure due to the seating, a result which agrees much better with experience.

Effective closing force of valves in terms of the velocity of the water.—Consider the conditions of equilibrium when the valve is open. Let the dimensions be taken in feet, the pressure in lbs. per sq. ft., the velocities in ft. per sec. Let G be the weight of a cubic foot of water (62·4 lbs.). Let q be the effective force tending to close the valve in lbs. per sq. ft. of its projected area. This consists of the weight of the valve in water (about seven-tenths of its weight in air) together with any applied closing force, divided by the area of the valve.

Since the volume of flow upwards through the passage under the valve at velocity v must be equal to that horizontally between valve and seat,

$$\pi d_1 h v_1 = \frac{\pi}{4} d^2 v \qquad . \qquad . \quad (1)$$

$$\frac{v_1}{v} = \frac{d^2}{4 d_1 h}.$$

But

$$v_1 = \sqrt{2g \frac{p - p_1}{G}}$$

$$= \sqrt{p - p_1} \qquad . \qquad . \qquad . \quad (2)$$

very nearly. Putting this value in (1),

$$d_1 h \sqrt{p - p_1} = \tfrac{1}{4} d^2 v.$$

But since there is equilibrium

$$p - p_1 = q \text{ lbs. per sq. ft.}$$

$$4 d_1 h \sqrt{q} = d^2 v$$

$$v = 4\frac{d_1 h}{d^2}\sqrt{q}.$$

Let $d_1 h/d^2 = k$, the ratio of area of opening of valve to area of passage below valve. Then

$$q = \frac{v^2}{16\,k^2}$$

$k =$	1	$\tfrac{3}{4}$	$\tfrac{1}{2}$	$\tfrac{1}{4}$	
$q =$	$\cdot 062\,v^2$	$\cdot 111\,v^2$	$\cdot 25\,v^2$	v^2	lbs. per sq. ft.

Taking $v = 3$ and 6 ft. per sec., and reducing to lbs. per sq. inch,

$k =$	1	$\tfrac{3}{4}$	$\tfrac{1}{2}$	$\tfrac{1}{4}$	
$v = 3$; $q =$	$\cdot 0039$	$\cdot 0069$	$\cdot 0156$	$\cdot 0625$	lbs. per sq. in.
$v = 6$; $q =$	$\cdot 0156$	$\cdot 0276$	$\cdot 0724$	$\cdot 2500$	,, ,,

which shows that the valve must be of small weight to give an area of lift equal to the passage under the valve.

This theory must, however, be taken only as a rough approximation. The form of the valve, affecting the curvature of the stream lines in the valve box, may greatly influence the lift for any given velocity of flow.

Some experiments by Prof. Bach [1] on small valves show that if Ω is the area through the valve seat, ω the area between the valve and seat, v the velocity through the valve seat, then the effective closing force is given by an equation of the form

$$q = \frac{v^2}{2g}\left[\alpha + \beta\left(\frac{\Omega}{\omega}\right)^2\right].$$

Let s be the length of stroke of the pump and n the number of strokes per minute; then for any given description of valve the limit of speed at which the valve closes quietly is given by an equation of the form

$$n^2 s = \text{constant},$$

[1] 'Versuche über Ventilbelastung,' Berlin, 1884; 'Zeitsch. d'Ingenieure,' 1886.

and at that limit the effective closing force is given by the equation

$$q = n^2 s \times \text{constant}.$$

133. *Prof. Riedler's valves.*—As accurate closing of valves is so important, and as the conditions to secure this

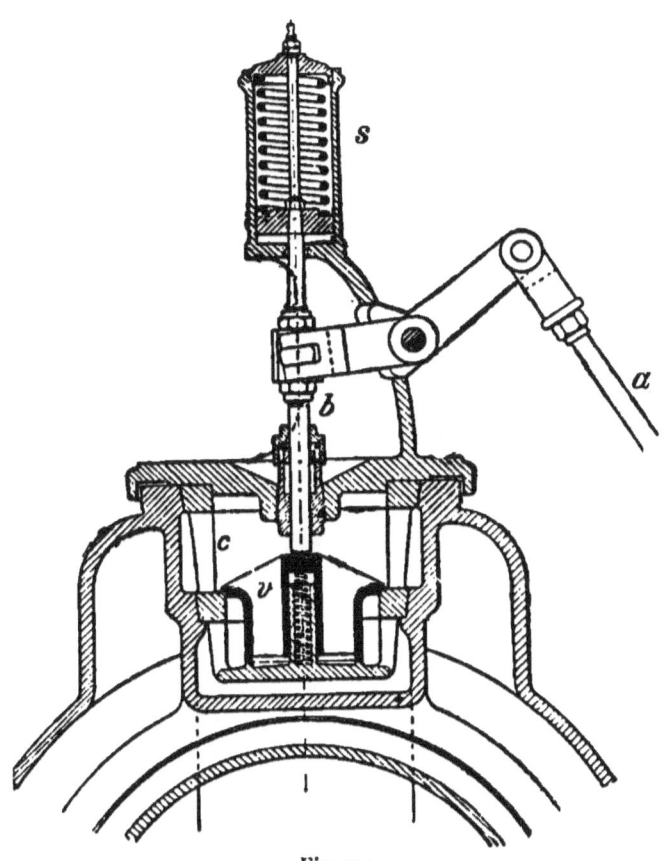

Fig. 134

conflict with the conditions of easy and full opening, it has occurred to Prof. Riedler to arrange valves so that they open very easily and automatically, and then to secure the closing at the proper moment by mechanism. He has applied such valves with very great success, both in pumping engines and air compressors.

Fig. 134 shows a valve for an air compressor. The valve v is a double-beat valve, and c is the valve seating. In the centre of the valve is a helical spring which presses the valve upwards, almost but not quite balancing its weight. Above the valve is a rod b, not attached to the valve, and driven by an eccentric or cam connected with the lever and rod d. This rod, except when lifted by the eccentric, is strongly pressed down by the spring s. The eccentric lifts the rod b before the time of opening of the valve, so that it opens by the action of the fluid pressure automatically. When the valve is to be closed, the eccentric allows b to force the valve down by the action of the spring s.

CHAPTER X

VALVE GEARS

134. *Steam-engine slide valves.*—Of all the kinds of valve used to effect the distribution of steam to steam-engine cylinders, the slide valve is by far the most commonly adopted. The full treatment of the action of the slide valve is beyond the scope of this treatise; a short description of the more simple slide-valve gears is all that can be attempted.

In its simplest form the slide valve consists of a dish-shaped rectangular piece (fig. 135), the face of which is accurately planed and sometimes scraped to a true plane surface. The valve slides upon a seating, also accurately planed, in the steam chest, and termed the *cylinder face* (§ 29). In the cylinder face are formed three *ports*, two communicating with passages leading to the ends of the cylinder, and termed *steam ports*, the third communicating with the atmosphere or condenser, and termed the *exhaust port*. The slide valve is pressed down on its seating by the excess of steam pressure on its back over that on its face, and leakage is prevented by the accuracy of fitting of the valve and its seating.

The ordinary form of the valve in longitudinal section is **D**-shaped, as shown in fig. 135. It has two flat faces, which, when the valve is in mid position, cover the steam ports, the arched part of the valve covering at the same time the exhaust port. If the valve moves in either direction from mid position it uncovers one steam port and admits steam to one end of the cylinder. At the same time the other steam port is put in communication with the exhaust passage. The reciprocating motion of the valve which opens the ports alternately is effected by an eccentric (p. 92), which may be regarded as a very short crank, keyed on the same shaft as the engine crank. It is obvious that the greatest travel of the valve each way from mid position is equal to the radius of the eccentric, except when modifying levers are interposed.

Slide valves have been made of cast iron, of gun-metal and of phosphor bronze. The wear of cast-iron valves appears to be one-third less than that of ordinary gun-metal valves, and the friction is less also. The faces of the valve and seating are sometimes scraped to true planes. It appears to be sufficient, however, in valves of moderate size, to plane the surfaces so that when in position the direction of planing for the valve is at right angles to that for the seating. They then wear to good surfaces.

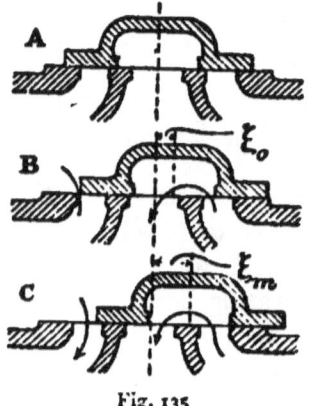

Fig. 135

Fig. 135 shows a slide valve in three positions. At A the valve is in mid position, the centre line of valve coinciding with the centre line of the cylinder face. At B the valve has already opened one port a little to steam and the other to exhaust. The valve has travelled a distance ξ_o from mid position. This is the position when the crank is at the dead point and the piston stroke beginning. The amount of the opening of the port at the moment the stroke of the engine begins is called the *lead* of the valve. At C the valve is at the end of its travel towards the right, and the valve has travelled a distance ξ_m from mid position equal to the radius of the eccentric.

Neglecting the small disturbance introduced by the

obliquity of the eccentric rod, the valve is at the ends of its travel when the eccentric is at its dead points on the line of stroke, and the valve is in mid position when the eccentric is 90° from its dead points, or at right angles to the line of stroke.

135. *Disturbance introduced by obliquity of eccentric rod. Setting valve to equalise lead.*—With an indefinitely long eccentric rod the travel of the valve would be exactly symmetrical on either side of mid position, equal travel on either side from mid position corresponding to equal angles of rotation of the crank and eccentric radius. Then, to secure a similar distribution of steam to both ends of the cylinder, the centre of the valve in mid position must coincide with the centre of the cylinder face. The obliquity of the eccentric rod introduces a small amount of unsymmetry in the forward and backward travel of the valve.

It is customary, in setting the valve in the workshop, to adjust the length of the valve rod so that the lead of the steam edge of the valve is the same for both steam ports; that is, the valve opening is made the same, with the crank at both dead points. It will then be found that the centre of the valve's travel does not exactly coincide with the centre of the cylinder face, being a little on the side of the interior dead point, or dead point nearest the cylinder.

136. *Lap and lead of slide valve.*—In the earliest slide valves the width of the faces of the valve was sensibly equal to the width of the steam ports. Then, the moment the valve passed its mid position, it began to open one port to steam and the other to exhaust. Apart from a circumstance to be mentioned presently, the steam piston must be at the end of its stroke at the moment steam begins to be admitted on one side and exhausted from the other. It follows that, with this form of valve, the valve is at mid stroke when the steam piston is at the end of its stroke; consequently the eccentric must be at right angles to the crank.

It was discovered, however, that with this arrangement the steam entered and left the cylinder with difficulty at the beginning of each stroke, in consequence of the very gradual opening of the slide valve. To afford a wider opening to the steam, it was found necessary that the valve should be already a little open at the beginning of a stroke. To secure this it is only necessary to fix the eccentric a little more than 90° in advance of the crank. The width of port open at the beginning of the stroke, at the steam edge of the valve, is termed the *lead*, and will be denoted by e.

Next it was found desirable to make the faces of the valve wider than the steam ports, so that when the valve is

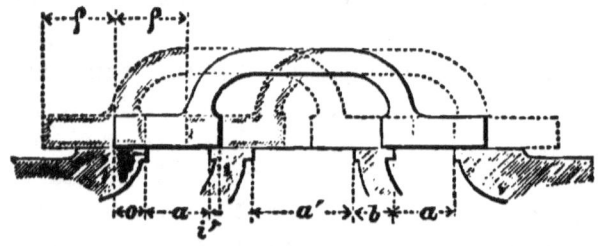

Fig. 136

in mid position (fig. 136) the valve faces overlap the edges of the ports. The width of overlap on the steam edge of the valve is called the *outside lap*, o, and that on the exhaust edge of the valve the *inside lap*, i. Generally the former is greater than the latter. Then one port opens sooner or more widely to exhaust than the other to steam, and this diminishes the back pressure without sensibly diminishing the work done by the steam in the forward stroke.

The valve must be open at the beginning of the stroke on the steam edge by the amount e fixed for the lead. Hence at the beginning of a stroke the valve must have already travelled from mid position a distance $\xi_0 = o + e$. The eccentric must therefore have also passed its mid position at right angles to the crank by an angle necessary to move the valve through a distance ξ_0. This angle, which will be denoted by θ, is called the *angle of advance*. The angle of advance may be regarded as made up of two angles, one corresponding to the movement of the valve through the distance o, which may be called the 'lap angle,' and one necessary to move it the further distance e, which is the opening of the port at the beginning of the stroke, and which may be termed the 'lead angle.' The whole angle between the crank and eccentric radius, or $90° + \theta$, which is a fixed angle for simple valve gears, may be called the *angle of keying*.

One effect of giving outside lap to the valve and advance to the eccentric is that the valve closes the steam port before the end of the forward stroke. Then the steam is cut off before the end of the stroke and the valve acts as an expansion valve. There is a limit to the amount of lap which can be used with an ordinary slide valve. To whatever extent the opening of the exhaust is made earlier, to the same extent the closing of the exhaust is made earlier also. If the exhaust is closed too soon, steam is retained in the cylinder and compressed, as the piston returns, into the small clearance space at the end of the cylinder. This action is termed 'cushioning.' A moderate amount of cushioning is useful, but excessive cushioning would be prejudicial. To prevent this the outside lap is usually not greater than is sufficient to close the steam port at $\frac{7}{8}$ths of the stroke. When more expansion is wanted, a double-slide valve or some other arrangement is used.

Fig. 136 shows a section of a slide valve and of the steam ports, taken parallel to the direction in which the valve moves. The dotted lines show the positions of the valve at the ends of its stroke, in either position completely uncovering one port to exhaust, and partially uncovering the other to steam. In this figure a is the width of steam port, o the outside lap, i the inside lap, and ρ the half travel of the valve or eccentric radius.

137. *Area of steam ports.* —The area of the steam ports must be so arranged that the mean velocity of the steam does not exceed 80 to 100 feet per second. Let Ω be the piston area and ω the area of each steam port, then the ratio of port area to piston area is as follows :—

Piston velocity feet per minute	$\dfrac{\omega}{\Omega}$
200	·04
300	·055
400	·07
600	·10

For locomotives which run at a high but variable speed, $\omega = ·07\Omega$. Let l be the length and a the width of each steam port. Then $al = \omega = \left(\dfrac{\omega}{\Omega}\right)\Omega$. The proportions of the port are variable; the length l may be from 0·5 to 0·8 of the cylinder diameter, and the ratio $\dfrac{l}{a}$ is 6 in small engines, 7 in medium engines, and 9 in large engines. Let D be the cylinder diameter, and let $l = x\,\text{D}$, $a = y\,\text{D}$. Then suitable values of x and y are given in the following table :—

Piston speed	x	y	x	y	x	y
200	·4	·078	·5	·062	·6	·052
300	·5	·086	·6	·072	·7	·062
400	·6	·091	·7	·078	·8	·068
600	·7	·112	·8	·098	·9	·087

It may be noted that the frictional loss of pressure of steam in steam passages is proportional to its density, and its density increases almost directly as its absolute pressure. Hence, if the loss in the passages is to be restricted to a given number of pounds pressure, the ports must be larger the greater the pressure of the steam. Some remarks on the unexpectedly large loss of pressure in passages and ports when high pressure steam is used will be found in Willans on Non-condensing Steam Engine Trials, 'Proc. Inst. Civil Eng.' vol. xciii. Some constructors allow about 90 sq. ins. of port opening to steam and 140 sq. ins. to exhaust, per lb. of steam used in the cylinder per stroke. This rule allows an area increasing directly as the density of the steam at release.

The whole width of the steam port is opened to exhaust, but often only 0·6 to 0·9 of the width to steam.

138. *Proportions of the slide valve* :—

Let $a=$ width of steam port.
  $na=$ greatest width opened to steam.
  $o=$ outside lap.
  $i=$ inside lap.
  $e=$ lead ; $e'=$ inside lead.
  $b=$ width of bar between steam and exhaust ports.
  $a'=$ width of exhaust port.
  $\rho=$ half travel of valve, or radius of eccentric.
  $r=$ radius of crank or half stroke of engine.
  $\iota=$ ratio of eccentric radius to length of eccentric rod.
  $\xi=$ distance valve has travelled from its mid position when the crank has moved through an angle ϕ from the dead point.
  $l=$ distance piston has travelled from beginning of stroke at the same moment.
  $\theta=$ angle of advance of eccentric, so that the eccentric is $90°+\theta$ in advance of the crank.

The width b of the bars is fixed empirically. In small engines it may be $=\frac{a}{2}+\frac{1}{4}$, and in large engines it is determined almost entirely with reference to convenience of casting, and should be at least equal to the cylinder thickness. The inside lap, i, is generally small, and may be from ·075a to ·1a. The outside lap, o, is determined by the point at which steam is to be cut off. Very commonly o is from $0{\cdot}25a$ in slow to $0{\cdot}66a$ in fast engines. Then, if the valve and eccentric are directly connected,

$$\rho = na + o = a + i \quad . \quad . \quad . \quad (1)$$

If these equations are satisfied, o and i are not independent when a and na are fixed.

The lead e may be from $\frac{1}{16}\rho$ in slow to $\frac{1}{8}\rho$ in fast engines.

The angular advance of the eccentric, θ, if the obliquity of the eccentric rod is neglected, is determined by the equation—

$$\sin \theta = \frac{o+e}{\rho} \text{ nearly} \quad . \quad . \quad . \quad (2)$$

This determines θ, and then the angle between the crank and eccentric radius, or angle of keying, is $90° + \theta$. The following table gives some values of θ :—

$\frac{e}{\rho} =$	When $\frac{o}{\rho} =$					
	·1	·15	·2	·25	·3	·35
$\frac{1}{16}$	9° 21'	12° 16'	18° 13'	18° 12'	21° 15	24° 22'
$\frac{1}{12}$	11 10	14 6	17 5	20 6	23 11	26 21
$\frac{1}{8}$	13 1	15 58	18 58	22 2	25 10	28 22

The equation is not quite exact, because of the obliquity of the eccentric rod, but the error is not great in ordinary cases.

The inside lead $= e' = \rho \sin \theta - i$. The inside and outside lead and lap are connected by the equation, $o + e = i + e'$.

The width a' of the exhaust port must be equal to or greater than $2\rho - b$.

The width of the hollow under the valve (measured parallel to the direction of the valves' motion) $= 2(a + b - \rho) + a'$.

139. *Travel of valve and corresponding crank angle when the influence of the obliquity of the eccentric rod is neglected.*— Let a line through the two dead centres of the crank-pin

circle be termed the 'line of stroke.' Generally this line is also parallel to the axis of the cylinder. If the obliquity of the eccentric rod is neglected, the valve is in its mid position when the eccentric radius is at right angles to the line of stroke. Let that position be termed, for shortness, the mid position of the eccentric. As the eccentric moves through an angle α from its mid position, the valve travels a distance

$$\xi = \rho \sin \alpha \quad . \quad . \quad . \quad . \quad (3)$$

which will be $+$ or $-$ according as α lies between $0°$ and $180°$, or between $180°$ and $360°$, α being measured in the direction of motion of the crank. Since the eccentric is $90° + \theta$ in advance of the crank,

$$\alpha = \phi + \theta,$$

where ϕ is the angle through which the crank has moved from its position at the beginning of the stroke. Hence

$$\xi = \rho \sin (\phi + \theta) \quad . \quad . \quad . \quad (4)$$

The opening of the port to steam is

$$w = \xi - o = \rho \sin (\phi + \theta) - o,$$

and the opening of the port to exhaust is

$$w' = -\xi - i = -\rho \sin (\phi + \theta) - i.$$

When admission begins and when steam is cut off, $w = 0$; and when exhaust or compression begins, $w' = 0$. Inserting these values, we obtain four values of the crank angle for each edge of the valve and for one revolution of the engine.

For			$(\phi + \theta)$ lies between
Admission	$w = 0$	$\sin (\phi_1 + \theta) = \dfrac{o}{\rho}$	$0°$ and $90°$
Cut off		$\sin (\phi_2 + \theta) = \dfrac{o}{\rho}$	$90°$ and $180°$
Release	$w' = 0$	$\sin (\phi_3 + \theta) = -\dfrac{i}{\rho}$	$180°$ and $270°$
Compression		$\sin (\phi_4 + \theta) = -\dfrac{i}{\rho}$	$270°$ and $360°$

From these equations the values of $\phi+\theta$, and therefore of ϕ, can be obtained. The angles are connected by the relations $\phi_2 = 180° - \phi_1 - 2\theta$; $\phi_4 = 180° - \phi_3 - 2\theta$.

The following form of the same equations is sometimes more convenient [1]:—

Since
$$\sin \theta = \frac{o+e}{\rho}, \quad o = \rho \sin \theta - e.$$

Hence,
$$w = \xi - o = e + \rho \left\{ \sin(\phi + \theta) - \sin \theta \right\}.$$

For admission and cut off, $w = 0$, and we get
$$\sin(\phi + \theta) = \sin \theta - \frac{e}{\rho};$$

For admission, $\phi_1 = 2\pi - \dfrac{e}{\rho \cos \theta}$;

For cut off, $\phi_2 = \pi - 2\theta + \dfrac{e}{\rho \cos \theta}$.

The angles are in circular measure in these equations, and can be reduced to degrees by multiplying by $\dfrac{180}{\pi}$ or by 57·3.

Similarly, since $i + e' = o + e$,
$$i = \rho \sin \theta - e$$
$$w' = -\xi - i = e' - \rho \left\{ \sin(\phi + \theta) + \sin \theta \right\}.$$

For release and compression, $w' = 0$; then
$$\sin(\phi + \theta) = -\sin \theta + \frac{e'}{\rho};$$

For release, $\phi_3 = \pi - \dfrac{e'}{\rho \cos \theta}$;

For compression, $\phi_4 = 2\pi - 2\theta + \dfrac{e'}{\rho \cos \theta}$.

[1] Resal, 'Mécanique Générale,' vol. iv. p. 238.

140. Position of piston for given crank angles, when the obliquity of the connecting rod is neglected.—If l is the distance the piston has travelled from the beginning of its stroke, when the crank has revolved through the angle ϕ, measured from the dead point, then if the obliquity of the connecting rod is neglected

$$l = r(1 - \cos\phi) \qquad . \qquad . \qquad . \qquad (5)$$

where $\cos\phi$ is negative, if ϕ lies between 90° and 270°. By inserting the values of ϕ, obtained above, we obtain approximately the position of the piston for admission, cut off, release and expansion. As, however, the obliquity of the connecting rod sensibly affects the position of the piston, it is better to set off the positions of the crank corresponding to the above values of ϕ on a diagram drawn to scale, and then by laying off the connecting-rod length, the position of the piston is found exactly; or the graphic constructions given below may be used.

141. Crank angles corresponding to given ratios of expansion.—Let l_2 be the travel of the piston corresponding to the crank angle ϕ_2 at cut off. Then $l_2/2r$ is the ratio of cut off. The following table gives the relation between these quantities, when the obliquity of the connecting rod is neglected:—

$\dfrac{l_2}{2r} = 0.4$	0.45	0.5	0.55	0.6	0.65
$\phi_2 = 78\frac{1}{2}°$	83	90	96	$101\frac{1}{2}$	$107\frac{1}{2}$
$\dfrac{l_2}{2r} = 0.7$	0.75	0.8	0.85	0.9	0.95
$\phi_2 = 113\frac{1}{2}°$	120	127	$134\frac{1}{2}$	143	154

The ratio $2r/l_2$ is the ratio of expansion or number of times the steam is expanded.

142. Graphic methods of determining the relation between piston travel and crank angle. Müller circles.—The alteration of the position of the piston due to the obliquity of the connecting rod cannot be neglected in studying valve gears. The algebraical expression for the piston travel, when the obliquity of the connecting rod is taken into account, is complicated, but there are easy ways of finding it graphically. Fig. 137 shows a construction due to Prof. Müller. Let A B be the cylinder, the piston being initially at A, P e the connecting rod, and e c the crank. For sim-

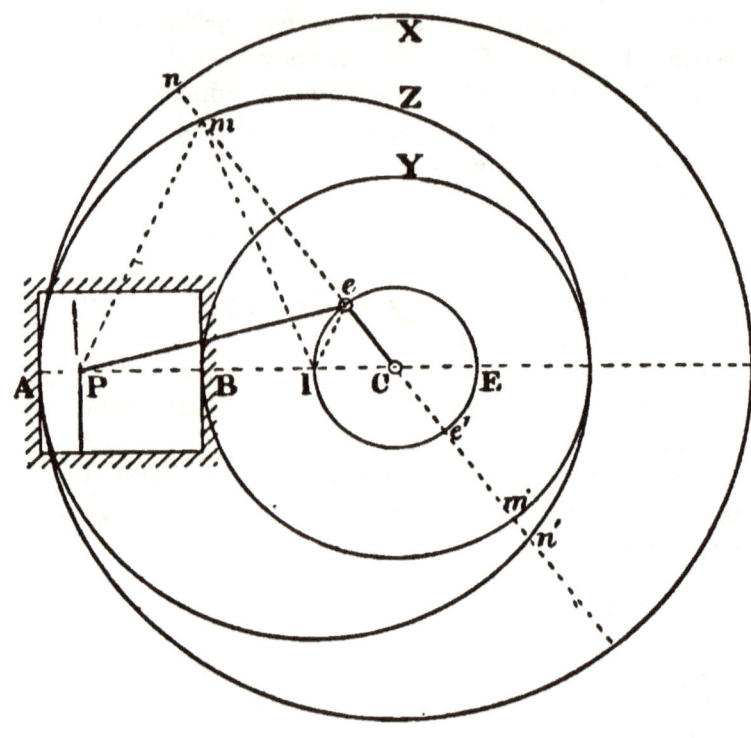

Fig. 137

plicity the piston is reduced to a line, and the connecting rod supposed attached directly to the piston, but this in no way alters the motion.

Let the crank length $e\,c = R$, the connecting-rod length $e\,P = L$. From centre c with radius $L + R$ describe the circle X and with radius $L - R$ the circle Y. From centre I with radius L describe the circle Z. The points I and E are the interior and exterior dead points. As the crank travels from C I to C e, the piston travels a distance A P, which may be found by taking e P = L. It is more easy in many cases to produce C e to cut the circles Z X in $m\,n$. Then $m\,n =$ A P. Join m I, m P, e I. In the triangles I $e\,m$, I e P, m I $=$ e P, being radii of the circle Z; and e I is common to the two triangles. Also since C $e =$ C I, the exterior angle $m\,e$ I $=$ P I e. Hence the triangles are superposable and $e\,m =$ I P. But $e\,n =$ I A. Therefore $m\,n =$ A P, the piston travel.

The Zeuner valve diagram gives directly the crank positions for given positions of the slide valve. If Müller circles are drawn to any convenient scale outside the valve diagram, the lengths of piston travel for any crank position are easily scaled off. Neglecting the eccentric obliquity which produces generally only a small effect, it will be found that if C e is a position of the crank (say at cut off) when the action of the steam on the left of the piston is considered, then C e' is the corresponding position of the crank when the action of the steam on the right of the piston is considered. If $m\,n$ is, for instance, the distance of the piston from A when cut off takes place on the left of the piston, $m'\,n'$ is the distance of the piston from B when cut off takes place on the right of the piston. It is due to the different obliquities of the connecting rod for these two positions that $m\,n$ is not generally equal to $m'\,n'$.

A still easier construction for finding the piston travel is the following:[1] Let C e, fig. 138, be the crank of radius R, and e P the connecting rod of length L as before. With centres A and B and radius equal to L, draw the arcs X I X, Y E Y, which may be termed the 'inner and outer dead point arcs.' If I E is taken to represent the stroke, we can find the point f corresponding to P by drawing the arc $c\,e\,f$ with

---

[1] Coste and Maniquet, 'Traité des Machines à Vapeur,' 1886. Grashof, 'Maschinenlehre,' vol. iii. 1890.

centre P and radius $Pe = L$. Since $AI = L$, then $If = AP$, the piston travel from the end of the stroke. If through e a line bg is drawn parallel to EI, then it is easy to see that $fI = eg$. For the crank-pin position e, the distance of the piston from mid stroke is $cf = ea$. If the connecting rod were infinitely long, eb would be the distance from mid stroke. Hence ab is the deviation of the piston position due to obliquity of connecting rod.

It is convenient to notice that if any line YX is taken parallel to the line of stroke, an arc through e parallel to

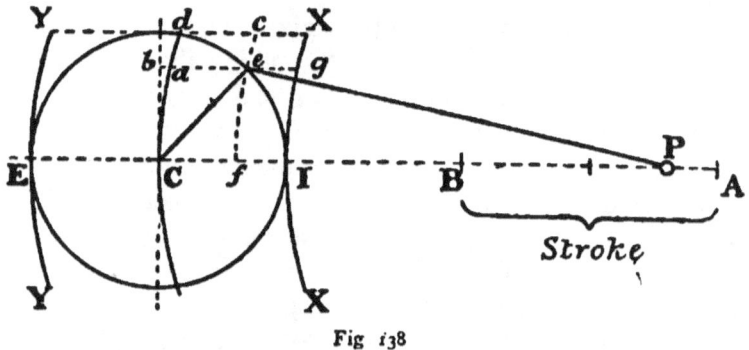

Fig 138

XX or YY cuts this in c, so that cd is the piston distance from mid stroke.

VALVE DIAGRAMS

143. *Graphic representation of valve action. Valve Ellipses.*—The direct calculation of the relative position of the valve and piston becomes very complicated if the obliquity of the connecting rod and eccentric rod are taken into account. Hence graphic methods are adopted which more or less completely evade the difficulty. Some of these are so far approximate in principle that, while they permit the obliquity of the connecting rod to be taken into account, they neglect the obliquity of the eccentric rod. All are, of course, necessarily approximate in this sense, that they

depend on the accuracy with which geometrical constructions can be made with rule and compass.

A very old and useful method of representing the action of a slide valve is to plot the motion of the piston as abscissa and that of the valve as ordinate. Then, if the obliquities of the rods are neglected, the curve obtained is an ellipse. In a simple way, however, the obliquities of the rods can be taken into account, and then the curve is no longer a true ellipse. Perhaps no graphic representation is more convenient than this for exhibiting the action of a valve gear already designed. It is less fruitful than some other diagrams in settling the proportions of the gear beforehand.

144. *Valve ellipse. Case I. Obliquity of rods neglected.*—Neglecting at first the obliquities of the rods, let 84, fig. 139, represent the stroke of the piston, and the circle described on 84 the crank-pin circle, the crank going round in the direction of the arrow. If we divide the crank-pin circle into parts 1, 2, 3 . . . . the corresponding positions of the piston will be found by dropping perpendiculars on 84. Now let the smaller circle represent the path of the eccentric centre. In order to project the valve movement at right angles to the piston movement the eccentric must be turned back through 90°. Hence, when the crank is at c 8, the eccentric radius will be c 8' the angle 8 c 8' being the angle of advance θ. Now, starting from 8', divide the eccentric circle into the same number of equal parts as the crank-pin circle, and number them to correspond. When the crank pin is at 1, the piston will have moved a distance 8 a from the dead point found by drawing 1 a vertically. At the same time the valve will be at a distance $a\,b$ from its middle position, found by drawing 1'b horizontally. b is a point in a curve the abscissa c a of which is the distance of the piston from mid stroke, and the ordinate $a\,b$ is the distance of the valve from mid position. This curve is the valve ellipse.

145. *Valve ellipse. Case II. Obliquity of both connecting*

234 *Machine Design*

rod and eccentric rod taken into account.—So far the diagram gives no more than can be found by simple calculations, though, as will be seen presently, the graphic representation

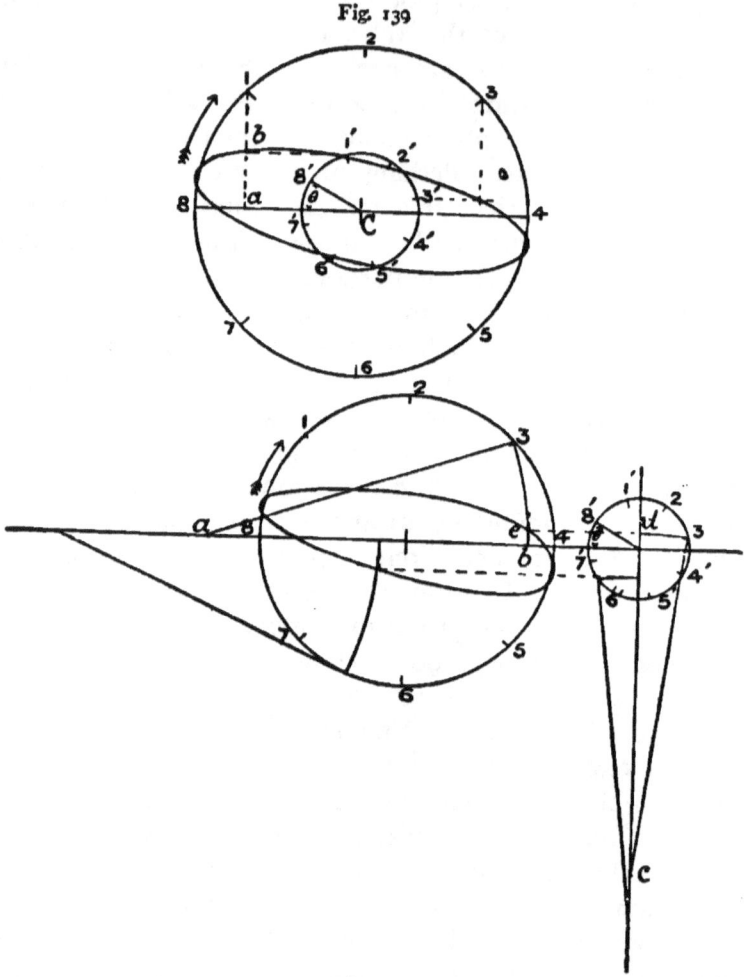

Fig. 139

Fig. 140

is a practically convenient one. It is easy, however, to take account of the obliquities of the rods, and then results are obtained which are calculated with difficulty.

Valve Gears 235

In fig. 140 the larger circle is the crank-pin circle and the smaller the eccentric circle, and they are divided exactly as in the last figure. Now, to find the exact position of the piston corresponding to any crank-pin position 3, it is only necessary to take $3a =$ the connecting-rod length and strike the arc $3b$ with radius $a\,3$. b is the piston position, allowing for the obliquity of the connecting rod. Similarly, to find the true position of the valve, it is only necessary to take $3'c =$ the eccentric-rod length (from centre of eccentric to centre of valve-rod pin). Striking an arc $3'd$ with radius $c\,3'$, d is the true position of the valve, its line of stroke being perpendicular to that of the piston. Projecting b vertically and d horizontally, we get a point e on a curve which gives the true relative motion of the piston and valve, and which may still for convenience be called the 'valve ellipse,' though it is not an exact ellipse.

It will be convenient generally to take a larger scale for the ecentric radius than for the crank radius, and this in no way renders the construction inexact.

Fig. 141 shows how the action of the valve may be rendered clear by the aid of a valve ellipse. In this figure a true ellipse has been drawn, but the construction is the

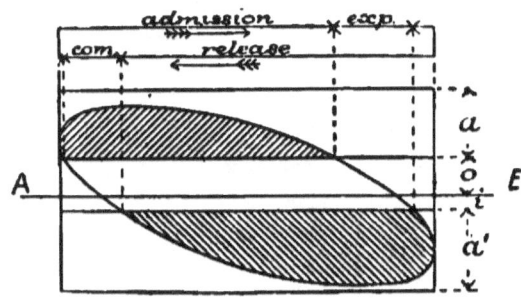

Fig. 141

same if an exact curve has been drawn, taking account of the obliquities of the rods. Let A B be the line of piston stroke corresponding to 84 in the previous figures. Draw lines parallel to A B at distances equal to the outside lap o, the inside lap i, the steam-port width a, and the exhaust-port width a' (see fig. 136). The shaded areas show at once the periods when the distance of the valve from mid position is greater than o or i, and these are the periods when the port is open to steam and exhaust to the left end of cylinder during one forward and return stroke. For the other end of the cylinder the lines parallel to A B must be reversed in position. Vertical lines permit the marking out of the periods of admission, expansion, release and compression.

If the eccentric rod is long compared with the eccentric radius, the piston positions may be determined as in Case II., and the valve positions more simply, as in Case I.

146. *Reuleaux, Reech, or Coste and Maniquet diagram.*—By using the same circle to represent (to different scales) the crank-pin circle and the circle described by the eccentric centre a very simple and useful diagram is obtained. When the obliquities of the rods are neglected, the diagram has long been known in France as Reech's diagram, and in Germany as Reuleaux's diagram. Lately, Coste and Maniquet have indicated how the diagram may be improved so as to take account of the obliquities of the rods.[1] It may be used very conveniently to obtain the data necessary for plotting valve ellipses. The exact diagram will be given first, and then the modification for the case when the eccentric rod is so long that the effect of its obliquity may be disregarded.

147. *Exact determination of the valve travel, taking account of the obliquity of the eccentric rod.*—Let the circle E B I, fig. 142, represent to one scale the crank-pin circle, I E being the piston stroke, and to another scale the circle of the eccentric, I E being then the total valve travel. O E $=R$ is the crank radius to one scale, and O B $=\rho$ is the eccentric radius to the other scale. Let θ be the angle of advance, so that when the crank pin is at E, the eccentric centre is at B. Bisect I E in O and draw A A at right angles.

[1] 'Traité des Machines à Vapeur,' Paris, 1886.

Valve Gears

With radius equal to the length of eccentric rod, or with a templet if the radius is inconveniently long, draw arcs X X, Y Y, touching the valve circle in I and E. These may be termed the 'interior and exterior eccentric dead-centre arcs.' Through o draw a parallel arc z z which may be termed the 'mid-travel arc.' Also, at a distance from o equal to the outside lap o, draw parallel arcs, which may be termed 'lap arcs.'

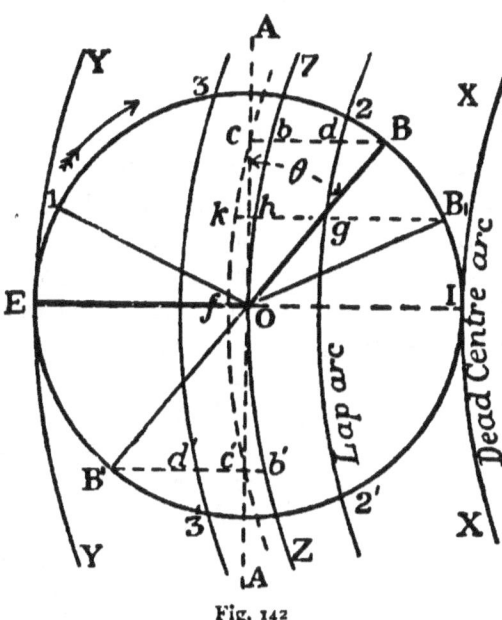

Fig. 142

When the crank is at O E the eccentric radius is at O B, and the travel of the valve from mid position is $\xi_o = B b$ exactly. The opening of the port is less than the valve travel by the amount of the outside lap o—that is, by the quantity $b d$. Hence d B is the opening of the port at the beginning of the stroke or the lead e.

When the crank has come to I, the eccentric will be at B', the valve travel will be $\xi_o' = B'b'$, and the port opening B'd'. Now, it is easy to see that, in consequence of the obliquity of the eccentric rod, $b'b'$ is greater than Bb, and the lead $B'd'$ greater than Bd, by the quantity $c'b'+cb$. It is customary in the workshop to adjust the valve to equal leads by lengthening the valve rod. The effect of this on the diagram is easily seen. Draw a new arc with the same radius through cc'. Then, if the valve is displaced so that its mid position is distant $u=of$ from the centre line of the valve face, the port opening to steam will be $\xi_0-o+u=Bd+bc$, and $\xi_0'-o-u=B'd'-c'b'$ for the right and left hand ports respectively, and these quantities are now equal.

If the crank comes to 1 and the eccentric to B_1, turning through equal angles, the valve travel will be B_1h and the port opening will be B_1g, if the valve is not adjusted, or B_1g+hk, if the valve is adjusted to equal leads. The quantity u is to be added to the valve travel for all positions of the eccentric to the right of A A and deducted for all to the left. If new lap arcs are drawn at a distance u to the left of the old ones, the port opening is the horizontal distance from any eccentric position B_1 to the new lap circle.

If the eccentric rod is indefinitely long the arcs x y z . . . become straight lines parallel to A A, and in that case no adjustment of the valve to equal leads is necessary.

148. *Complete valve diagram, drawn both for a long and short eccentric rod. Determination, also, of piston positions.*—Figs. 143 and 144 show complete valve diagrams, fig. 143 being drawn for the case where the eccentric rod is very long compared with the eccentric radius; fig. 144 for the case where the eccentric rod is exceptionally short.

Let E A B 1 be the crank-pin circle, E and 1 being the dead points. Through 1 and E draw arcs touching the circle with radius equal to the connecting-rod length. Then for any position o 1 of the crank the piston will be at a distance 1 m from one end of its stroke and 1 n from the other, *exactly*. The piston positions are thus given in terms of the crank positions.

Now, on this diagram the valve diagram shown in fig.

Valve Gears

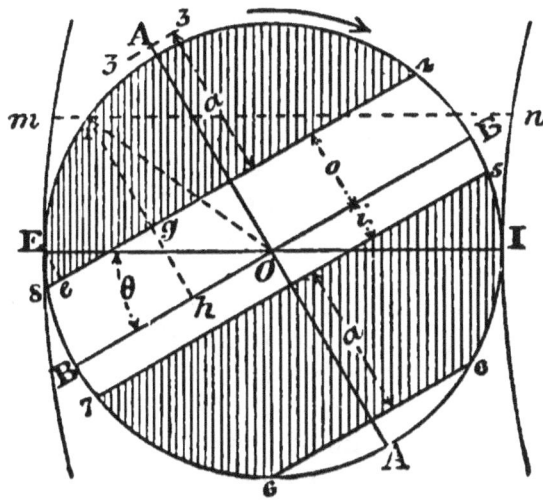

Fig. 143

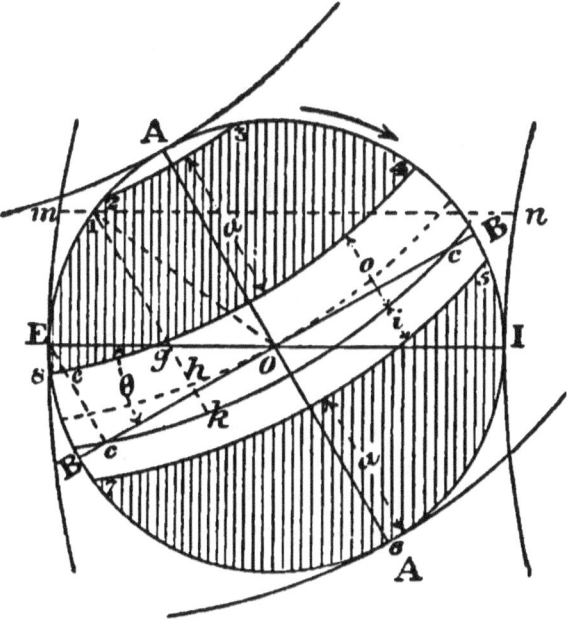

Fig. 144

142 is to be superposed, such scales being chosen that O E represents both the crank and eccentric radius. It is convenient to turn back the eccentric, fig. 142, through an angle $90° + \theta$, so that O B may coincide with O E, and the piston and valve travel may be measured from the same point of the valve circle. In fig. 143 draw B B, making an angle θ equal to the angle of advance with E I. Then the eccentric radius having been turned back $90° + \theta$ relatively to the direction of rotation coincides with the crank, and the valve travel is to be measured from the crank position at right angles to B B. Draw A O A at right angles to B B. Draw 48 and 57 parallel to B B at distances equal to the outside lap o and inside lap i. Draw 66 and 33 at a further distance equal to a, the width of port.

For any crank position, O I, the whole valve travel ξ from mid position will be I h, measured parallel to A A, and the opening of the port to steam will be $\xi - o = \mathrm{I}\, g$. If the crank position is taken in the lower semicircle we get the opening of port to exhaust in the return stroke.

The valve opens to steam at 8, and E e parallel to A A is the lead. The port remains open till the crank reaches 4, when steam is cut off. The valve, in this case, travels a little short of the port edge 33 at the extreme travel. At 5 the exhaust edge opens and steam is released. The port remains wide open to exhaust as the crank passes from 6 to 6, and compression begins when the crank is at 7. The corresponding positions of the piston can be found by the distances of the points on the crank-pin circle from the dead-centre arcs.

Fig. 144 is the same diagram with the correction necessary if the eccentric rod is short and if the valve is adjusted to equal leads. At A A draw arcs touching the valve circle with radii equal to the eccentric-rod length. Draw B B as before, making the angle θ with E I. From E drop a perpendicular E c on B B. Through c draw an arc parallel to the arcs at A A. This arc will cut the valve circle in those positions of the crank pin in which the centre of valve and centre of cylinder face coincide, after the valve has been adjusted to equal leads. From the arc through cc, parallel to A A, measure distances equal to the outside and inside laps and draw the lap arcs parallel to the arcs at A A.

The port opens to steam when the crank is at 8 and Ee is the lead. The port is fully open with the crank at 2, and begins to close with the crank at 3; cut off occurs with the crank at 4; release begins with the crank at 5; the port is just fully open to exhaust with the crank at 6, and compression begins with the crank at 7. The widths of the shaded parts measured parallel to A A are the widths of port open.

To study the action of the steam in the other end of the cylinder it is only necessary to set off the laps o and i in reversed positions relatively to the arc through cc and then to draw the lap arcs.

Problems.—Given θ, ρ, o and i. Then the valve diagram is obtained precisely as described above.

Given θ, ρ, e and the lead to exhaust e'. From E with e as radius describe a circle. Draw 84 (fig. 143) or the arc 84 (fig. 144), touching this circle and parallel to B B or the arc through c. With centre I and radius e' describe a circle, and draw 57 or the arc 57 touching this and parallel to B B or the arc through c. The inside and outside laps are then determined.

Given ρ, e, and the fraction of the stroke at which steam is to be cut off. To find the outside lap and angle of advance. With centre E and radius e describe a circle. At the given fraction of the stroke measured from E towards I draw an arc parallel to the arc at I. This will cut the valve circle in the crank position 4, at which steam is to be cut off. Draw 48 parallel to B B and touching the circle described round E.

Given the fraction of the stroke at which steam is to be released, together with o and θ to find the inside lap.

Along E 1 set off the distance the piston moves before release takes place. Through the piston position at release draw an arc parallel to the arc at 1. This gives 5, the crank position at release. Draw 57 parallel to B B; this determines i.

Given ρ, e, and the fractions of the stroke at which cut off and compression occur. To determine θ and the inside and outside lap. With centre E and radius e describe a circle. Setting off along E 1 the lengths of stroke which correspond to admission and compression, draw arcs parallel to the arc through 1. These determine the crank positions 4 at cut off and 7 at compression. Draw 48, touching the circle round E. This determines the angle of advance and outside lap. Draw 75 parallel to 48; this determines the inside lap.

149. *Zeuner's polar valve diagram for a simple valve.*[1]— The valve diagram of Prof. Zeuner, of Dresden, is one of the simplest, and it permits the obliquity of the connecting rod to be taken into account without much complexity, though not the obliquity of the eccentric rod. As it consists entirely of circles it is easily drawn, and gives a very approximate solution of most practical problems on valve gears by easy constructions. The principle is this: The travel of the valve from the middle of its stroke is given directly for any crank position; thence the corresponding piston position can be found.

Let fig. 145 represent the mechanism of an engine, which for definiteness is supposed horizontal. ob, bc are the positions of the crank and connecting rod, and of, fg the corresponding positions of the eccentric radius and eccentric rod. The crank is supposed moving in the direction of the arrow. By setting off from $a_0 a_1$ lengths equal to bc, the piston stroke can be marked out. By setting off from $s_0 s_1$ lengths equal to fg, the valve stroke

---

[1] For fuller information consult Zeuner's 'Treatise on Valve Gears,' translated by Professor Klein.

Valve Gears 243

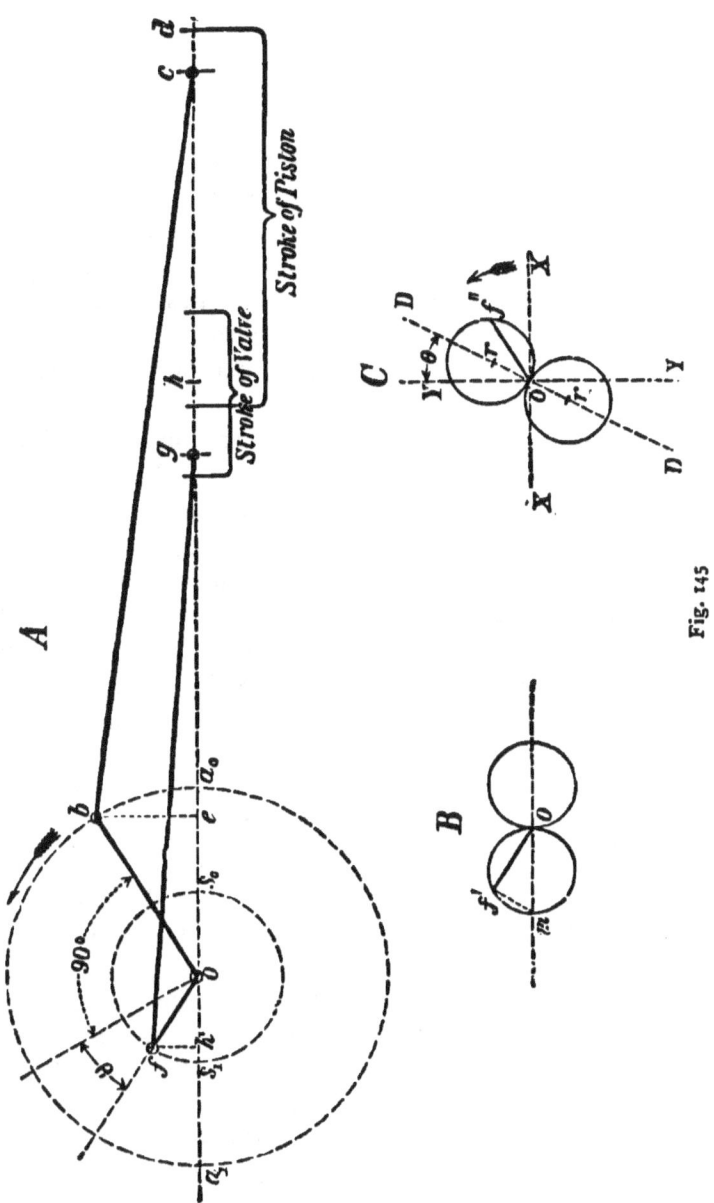

Fig. 145

can be marked out. Further, it may be noted that the angle bof is equal to $90°$ + the angle of advance, or $90° + \theta$.

150. *Theorem I. The polar locus of the valve travel reckoned from mid stroke is a pair of circles.*—Draw be, fk perpendicular to the line of stroke, and bisect the valve stroke in h. Confining attention at present to the movement of the valve, the valve travel, reckoned from mid stroke, in the position of the mechanism shown, is $\xi = hg$. Since $oh = fg$, and (when the eccentric rod is long compared with the eccentric radius) $kg = fg$ nearly, therefore $ok = gh$ nearly. In Zeuner's diagram the quantity ok is taken as a sufficiently accurate approximation to the valve travel ξ.

Now in diagram B, with centres on any horizontal line, and radii equal to half the eccentric radius, draw the two valve circles touching at o. Draw of' parallel to of in diagram A and join $f'm$. In the triangles ofk, $of'm$, $om = of$, the angle $mof' = kof$, and the right angle $of'm = okf$. Hence $of' = ok$. Therefore the diagram B will answer as a valve diagram, because the radius vector of' drawn parallel to any position of the eccentric will be approximately enough equal to the travel of the valve from its mid position; and knowing the travel of the valve, it will be easy to infer the condition of opening of the ports to steam and exhaust. The diagram B, however, will be more convenient if it is rotated backwards through an angle $90° + \theta$, as shown at c. Then the valve travel of', which in diagram B is parallel to the eccentric radius in A, comes to of'' in diagram c parallel to the crank ob in diagram A.

The diagram c is the simplest form of Zeuner's diagram. Two circles are drawn which are termed *valve circles*, the diameters of which are each half the total travel of the valve, and which touch at o. The centres of the valve circles are on a line which makes an angle $90° - \theta$ with the crank at the beginning of the stroke. *Then if a line of'' is drawn parallel to any position of the crank of the engine, the intercept*

of' is the travel ξ of the valve reckoned from the middle of its stroke, for that position of the crank. To draw the valve diagram C correctly proceed thus: Take any rectangular axes XX, YY, the former being parallel to the line of stroke. From YY set off the angle of advance θ *towards the initial position of the crank*. On the line DD so obtained take orl $or=$ half the eccentric radius. Then rr are the centres of the valve circles. Lastly, the radius vector of the valve circles, drawn from o parallel to any position of the crank, is the corresponding travel ξ of the valve from its mid position.

Generally it is necessary to find the piston positions also, and this can easily be done when the crank position is known. In cases where the obliquity of the connecting rod may be neglected, it is only necessary to drop a perpendicular be, diagram A, on the line of stroke, then a_0e will be the travel of the piston while the crank pin moves from a_0 to b. If to any scale a_0a_1 represents the whole piston stroke, e will be the piston position when ob is the crank position. Neglecting the obliquity of the connecting rod, however, introduces a not inconsiderable error. The exact piston travel can be ascertained by either of the methods in § 142.[1]

151. *Zeuner's valve diagram for a simple valve with lap circles.*—In fig. 146 let I E, Y_0Y_1 be the rectangular axes, the motion of the crank being from I to E in the direction of the arrow. Draw D_0D_1 making the angle of advance θ with Y_0Y_1 on the side of I. Take OD_0, OD_1, each equal to the half travel of the valve or to the eccentric radius, and on these lines as diameters describe the valve circles. Take OI equal on any scale to the crank radius, and draw the crank circle I 2 3 E. With centre o and radius equal to the outside lap o, draw the outside lap circle aa;

---

[1] The geometry of Zeuner's diagram and the working out geometrically of a number of problems is given in a treatise by Mr. Cowling Welch on 'Designing Valve Gearing.'

with centre o and radius equal to the inside lap i, draw the inside lap circle bb.

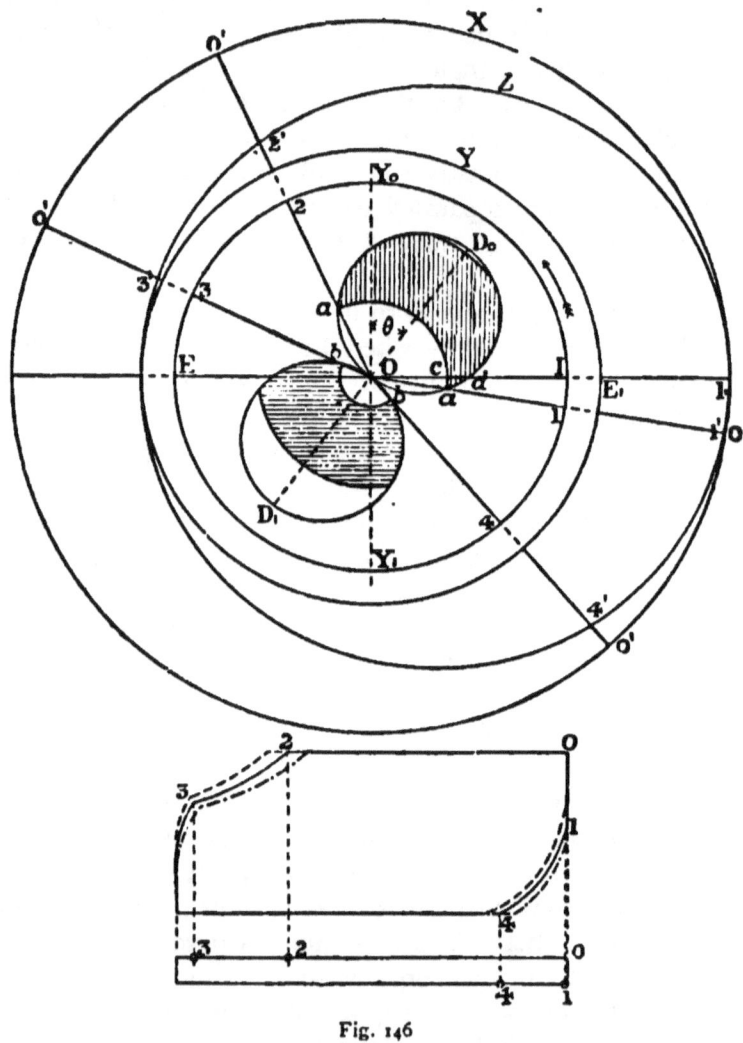

Fig. 146

The port opens to steam when the valve has travelled a distance equal to the outside lap, so that $\xi = o$. Hence $oa\,1$ is the position of the crank when the valve opens. At the beginning of the stroke, when the crank is at o 1, the travel of the valve is o d. Hence $cd = \xi - o$ is the lead e. The valve closes to steam when ξ is again diminished to o. Hence o a 2 is the position of the crank when the valve closes and expansion begins. Similarly o b 3 is the position of the crank when the valve opens to exhaust, and o b 4 that when the valve closes to exhaust and compression begins.

By drawing circles with centres at o and radii equal to $a + o$ and $a + i$, where a is the width of port, we mark out the periods during which the valve is fully open to steam and to exhaust. In the figure the steam port width has purposely been taken narrower than is usual, and the opening of the port is shown by the width along any vector of the shaded areas. More commonly the width of port is at least such that the shaded area extends to D_0. It should be noted that $D_0 D_1$ bisects the angles aoa and bob.

The lower figure is a horizontal projection of the points determined by the valve diagram on the crank-pin circle. It gives the proportionate lengths of stroke for each period, if the obliquity of the connecting rod is neglected. Thus, o 2 is the period of admission; 2 3 the period of expansion; 3 4 the period of exhaust; 4 1 the period of compression; and 1 0 the period of preadmission.

The point 2 is the point of cut off, 3 that of release, and 4 that of cushioning. Round the valve diagram Müller circles have been drawn, $1'E'$ being the stroke (§ 142). Then radial distances such as $2'o'$ between z and x give the true piston travel, taking into account the obliquity of the connecting rod. With these distances the corrected indicator diagrams for the forward and return stroke have been drawn, and are shown by dotted lines in the figure below.

Suppose that in designing a valve gear there are given the ratio of cut off z, the eccentric radius or half valve travel ρ, and the lead e. In fig. 147 take $x_0 x_1$ parallel to the line of stroke. With radius ρ describe the circle D K A.

Find the position O K 2 of the crank at which expansion begins. This is found approximately by taking $\frac{BC}{AC} = z$, or more exactly by the method above. Join AK, and take KE, equal to the given lead e. Bisect AE in F. Then AF = FE is the necessary outside lap o. Take OG = FK, and draw

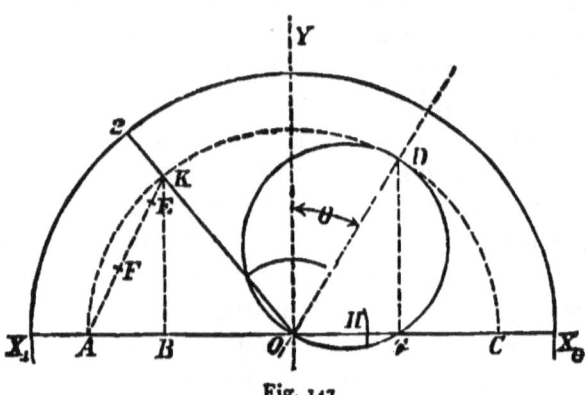

Fig. 147

GD perpendicular to XX, meeting the circle CDKA in D. Then DOY is the required angle of advance. On OD describe the valve circle. Take OH = FE, and through H draw the lap circle. The valve diagram can then be completed by drawing the other valve circle and the inside lap circle.

152. Construction of valves.—Fig. 148 shows a locomotive slide valve of the simplest form; the valve rod passes through the valve, the position of which is fixed in setting the valve by a pair of lock nuts at each end. The chief fault of the short D slide valve here shown is that the steam passages to the cylinder ends are necessarily long, and this increases considerably the clearance in the cylinders. The valve chest and cylinder face for such a valve are shown in fig. 26.

A considerably diminished clearance is obtained if the slide valve is divided, as shown in fig. 149.

The figure shows the valve chest of a jacketed engine. aa are the two separate portions of the D valve, bb are the two exhaust ports.

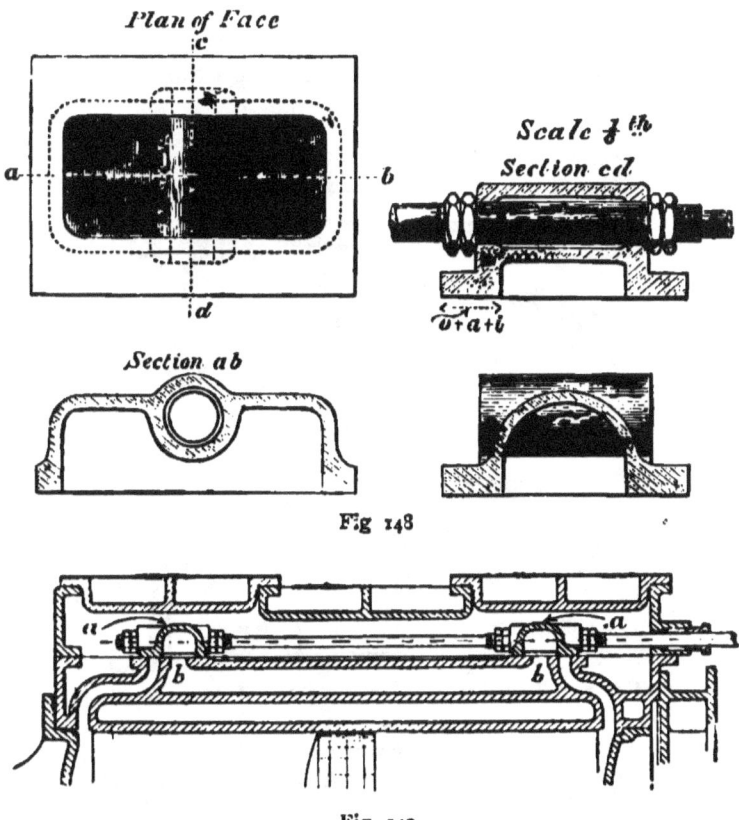

Fig. 148

Fig. 149

153. *The Trick valve.*—Fig. 150 shows a modified slide valve, giving a quicker and fuller opening of the steam port. A passage way is formed through the valve itself. This passage way overlaps the ends of the cylinder face at the ends of the travel so that steam is admitted not only at the steam edge of the valve but also through the valve passage. Supposing the valve in the figure to move to the right, steam will enter the left port, both at the left steam edge of the valve and through the passage which overlaps the right-hand end of the cylinder face. The pressure on the back of the valve when there is steam pressure inside the passage is diminished and the friction of the valve is reduced.

Fig. 151 shows a Trick valve at A in mid position; at B open to the extent of the lead at the beginning of a stroke,

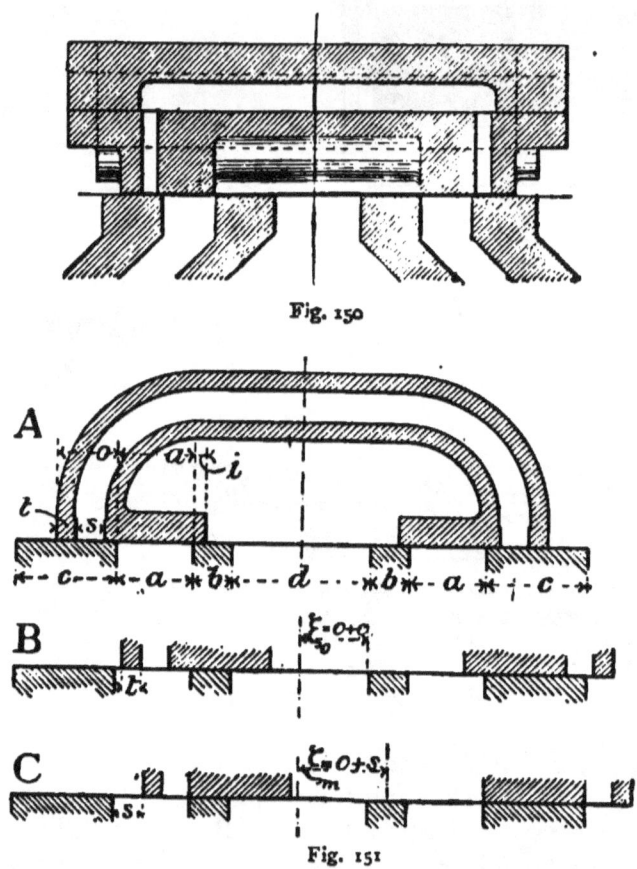

Fig. 150

Fig. 151

steam entering the left-hand port; at C at the end of its travel towards the right, the left-hand port being fully open to steam and the right-hand to exhaust.

Proportions of a Trick valve.—Let a be the width of port,

o the outside and i the inside, lap as indicated in fig. 150 a. The width t of the wall of steam passage must be taken arbitrarily. It is often $\frac{1}{2}$ to 1 inch. The width of bar between steam and exhaust port must also be taken arbitrarily; it is often taken $a/2$, or in large valves of the dimension easiest to cast.

It is convenient in a Trick valve that the passages should just completely open to steam and exhaust, as shown in fig. 151, C, where the left-hand port is just full open to steam and the right-hand port to exhaust. The width of steam passage through the valve is taken

$$s = \tfrac{1}{2}(a - t).$$

Then the eccentric radius or half travel is

$$\rho = a + i = o + s,$$

and o and i are not independent. Suppose

$$o = a + i - s.$$

The whole width of valve face from the exhaust edge to the extreme steam edge is $a + o + i$.

When the valve is at the end of its travel towards the right, there must still be an opening at least equal to a into the exhaust port. Hence, looking at fig. 151, C,

$$d + b + a - s = a + (a + o + i)$$
$$d = a + o + i + s - b.$$

Lastly, the width of cylinder face c beyond the steam port must be

$$c = 2o - t.$$

The width of the central aperture in the valve for exhaust is then

$$d + 2b - 2i.$$

The figure shows at A the valve in mid position. At B it has travelled to the right a distance $\xi_0 = o + e$, and the steam port on the left and the steam passage through the valve on the right are open to steam by the amount e, which in this case is therefore half the effective lead. At C the valve has travelled $\xi_m = a + i = o + s$, and both ports are open fully.

The valve diagram is drawn precisely as for a simple valve, merely remembering that the effective width of port is a to exhaust and $2s$ to steam.

154. *Balanced slide valves.*—The friction of the slide valve on its seating involves a considerable waste of energy, and this in very large valves becomes a serious loss. To reduce it double-ported valves are used, which, having only half the travel, involve only about half the loss of work in friction. This, however, is only a partial remedy. It is, perhaps an, even more prejudicial effect that the faces of the valve and valve seating wear away. The slide valve can be refaced when worn and the seating can be remade by adding a false seating. Nothing wears better than a cast-iron valve on a cast-iron seating, but then the wear is equal on both and both require refacing. Sometimes the valve is of gun-metal, or a gun-metal seating is fixed initially on the cylinder face, so that, the wear being chiefly concentrated on one face, the readjustment is easier.

A more complete remedy for the evils of valve friction is to use a balanced valve; that is, a valve so arranged that the pressure on the back is diminished. A rectangular or circular steam-tight space is formed between the back of the valve and the steam-chest cover, and this is put in communication with the exhaust. This relieves the valve from a great part of the pressure on its back.

Fig. 152 shows one form of balanced valve used in locomotives.

A rectangular space on the back of the valve is enclosed by a metal wall. In a groove in this wall are four cast-iron packing strips pressed upwards by springs. These strips slide on the planed face of a plate attached to the valve-chest cover. A small hole in the valve puts the space thus enclosed in communication with the exhaust. Hence on the enclosed area of the back of the valve the steam pressure

Valve Gears

acts only on the fixed plate and the valve carries only the exhaust pressure. The steam pressure is sometimes admitted below the packing strips. The steam pressure also tends to

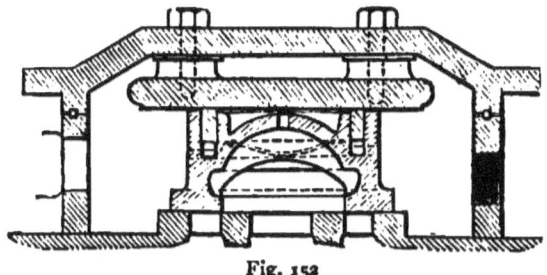

Fig. 152

force the strips against the inside of the groove and against each other.

Great care must be taken that there is no leakage of steam direct to the exhaust, or the waste of steam may be very serious.

CHAPTER XI

EXPANSION VALVES AND LINK MOTIONS

155. As in the case of simple slide valves it was necessary to confine attention to the most ordinary forms, so in dealing with the very extensive subject of Expansion Gears and Reversing Gears it will be necessary to select only some much-used types.

It was not uncommon at one time, especially in marine engines, to attempt to secure an earlier cut off than was possible with a single slide valve, by placing a second steam chest above the ordinary steam chest, with an expansion valve moved by a separate eccentric. As this second valve only regulated the admission of steam, it could be arranged to cut off as early as desired without entailing the evil of excessive compression during exhaust. But with this arrangement the whole steam chest containing the main or distributing valve formed part of the clearance volume of the cylinder up to the moment at which the main valve closed the cylinder port. This large clearance space almost nullified the action of the separate cut-off valve.

Since the steam in the clearance expands with tnat in the cylinder, it is obviously important, in engines intended to work with an early cut off, to reduce the clearance as much as possible. This is fairly well accomplished by putting the expansion valve on the back of the main valve, so that the addition to the clearance space, between the closing of the expansion valve and the closing of the main or distributing valve, is only the volume of the short passage through the main valve.

Fig. 153 shows an ordinary slide valve, in its central position, over the cylinder face. This valve has been

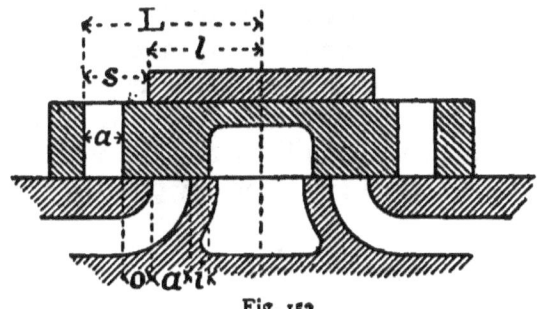

Fig. 153

extended at the ends and ports formed through it, usually of the same width as the ports in the cylinder face. On the back of this is an expansion plate driven by a separate eccentric. The expansion plate will cut off steam at the left port, if it travels a distance s to the left, relatively to the main valve; that is, if its centre line passes to the left of the centre line of the main valve by a distance s.

156. *Action of an expansion plate cutting off steam at its outside edges.*—Let o 1, fig. 154, be the crank at its inner dead point; let oa be the main valve eccentric of radius r_1 and angle of advance θ_1; and let ob be the expansion eccentric of radius r_2 and angle of advance θ_2. Join ab, draw oc equal and parallel to ab, and drop perpendiculars ad, be, cf on 1 E.

In the position shown, neglecting the eccentric-rod obliquity, the main valve will be at a distance o$d = \xi_1$ to the left of its mid position and the expansion plate at a distance o$e = \xi_2$. Consequently the relative travel, or distance of centre of expansion plate to the left of centre of main valve, is $\xi_2 - \xi_1 = ed$. This is the projection of ab on 1 E. But oc is equal and parallel to ab, hence o$f = de$. Now suppose

oc is a third eccentric fixed on the same shaft as the others and revolving with them. In all positions, $oc = \rho$ will be equal and parallel to ab. In all positions, the projection of oc on IE will be the difference of the projections of oa and ob, that is, to the distance apart of the plate and valve centres. Hence the relative motion is the same as if the main valve were at rest and the expansion plate driven by an eccentric, oc. Hence oc may be termed the *relative eccentric* or virtual eccentric of the relative motion. This relative eccentric has the radius ρ and the angle of advance ϕ. The relative travel $\zeta = of$ for any position will be equal

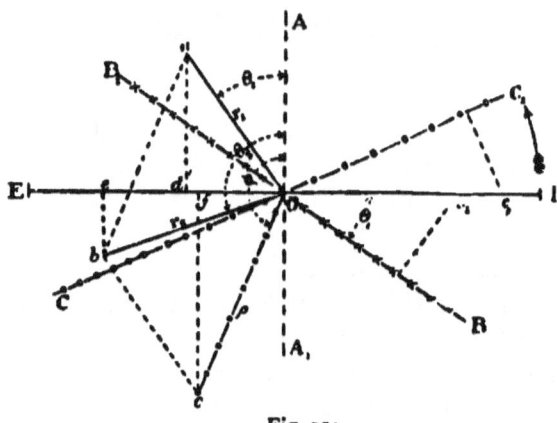

Fig. 154

to the perpendicular from c on AA_1, and will be a movement of the expansion plate to the left relatively to the main valve if c is to the left of AA_1, to the right if c is on the right of AA_1. When oc coincides with AA_1 the centre lines of the valve and plate coincide.

If, as in fig. 145, we turn backwards the valve circle for oa through an angle $90° + \theta_1$, then a will come to a_1 on the crank and AA_1 to BB_1. The travel of main valve from mid position will now be the perpendicular from a_1 on BB_1 and will be a travel to the left if a_1 is above BB_1. Similarly, if the relative eccentric circle is turned back through an angle

$90° + \phi$, c will come to the crank at c_1 and AA_1 to CC_1. The relative travel of expansion plate will be the perpendicular from c_1 on CC, and will be a travel to the left if c_1 is below CC_1.

In fig. 155, with crank radius R draw the crank circle having its internal and external dead points at I, E. At I and E with radius equal to the connecting-rod length draw the dead-centre arcs (see fig. 138). With radius $oa_1 = r_1$ draw the main valve circle, and with radius $oc_3 = \rho$ draw the

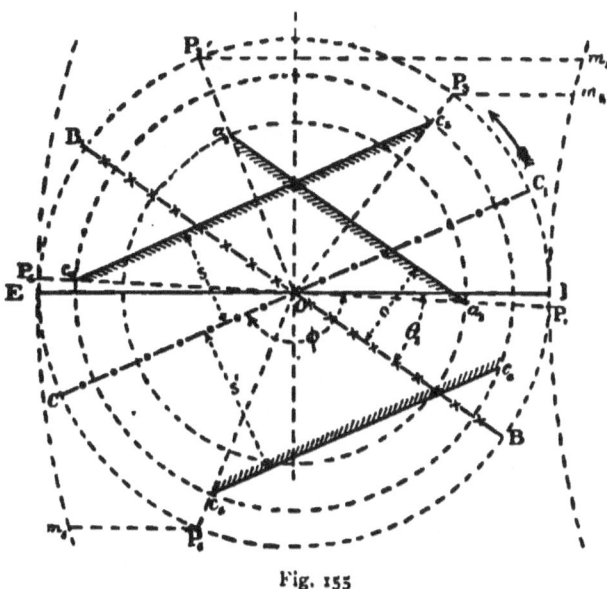

Fig. 155

relative motion circle. Suppose the directions BB_1 and CC_1 determined as in the previous figure; that is, BB_1 makes an angle θ_1, and CC_1 an angle ϕ, reckoned backwards to the direction of motion from I E. Draw a_1 a_2 parallel to BB_1, at a distance equal to o, the outside lap of the main valve, and draw c_3 c_4 parallel to CC_1 at a distance equal to s in fig. 153.

The main valve will open the right-hand port when the crank is at OP_1, because at that moment the travel of the main valve to the left will be the perpendicular from a_1 on BB_1, which by construction is equal to the outside lap. It will close the port again when the crank is at OP_2, because then the diminishing travel from mid position, the perpendicular from a_2 on BB_1, is again equal to the outside lap. The piston travel when steam is cut off will be $P_2 m_2$ drawn parallel to I E.

Now consider the action of the expansion plate. It has already been shown that the passage through the main valve will be closed when the relative travel is equal to s. But when the crank is at OP_3 the relative travel is the perpendicular from c_3 on CC_1, which by construction is equal to s. The piston travel when the expansion plate cuts off will be $P_3 m_3$. The expansion plate will reopen the port through the main valve when the crank is at OP_4, the relative travel being again equal to s. It is obvious that P_4 must fall later in the stroke than P_2 where the main valve cuts off, or steam will be a second time in the stroke admitted to the cylinder.

Suppose it is required that steam should be cut off at the same fraction of the forward and return stroke. Take $P_5 m_5$ equal to $P_3 m_3$. Then OP_5 is the crank position at cut off in the return stroke. Draw $c_5 c_6$ parallel to CC_1. Then the distance s' of $c_5 c_6$ from CC_1 is the value of s at the left-hand port necessary to give the same cut off as in the forward stroke. It is obvious that, in consequence of the obliquity of the connecting rod, the centre of the expansion plate must be displaced a distance $\frac{1}{2}(s'-s)$ by lengthening the valve rod if the cut off is to be equalised for both strokes.

157. *Meyer variable expansion gear*.—The action of the much-used Meyer expansion gear is easily derived from that of the expansion plate, already treated. Fig. 156 shows the arrangement of this gear. There is an ordinary slide valve, which is extended in length, and two admission ports formed in it equal in width to the ports in the cylinder face. This valve is driven by an eccentric in the ordinary way and completely controls the exhaust from the cylinder. The back of this valve is formed into a valve face, and a pair of expansion plates slide on this driven by a second eccentric. The function of these is to cut off the steam at any desired point of the stroke by closing the ports through the main valve. The expansion plates are connected by a right and left hand screw, which can be rotated by a handwheel outside the steam chest. By rotating this screw the distance

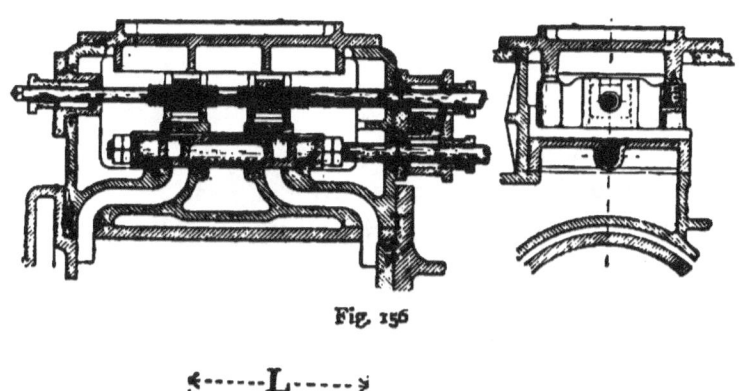

Fig. 156

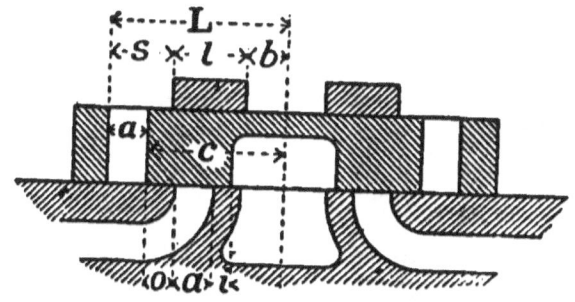

Fig. 157

between the expansion plates and the length between the cutting-off edges can be altered, the centre line between the plates keeping the same position. As the distance between the plates is increased cut off occurs earlier in the stroke.

Fig. 157 is a diagram of the valve and expansion plates in mid position. Let ρ be the radius of relative eccentric; then the greatest travel of the expansion plates each way from the centre of the main valve is ρ.

Let b_1 and b_2 be the greatest and least values of the half distance between the plates, and s_1, s_2 the corresponding values off s. Then b_1 and s_1 will correspond to the earliest cut off and b_2 and s_2 to the latest. In order that the expansion plate may not reopen the port at its inside edge, which is most likely to occur when the plates are furthest apart,

$$b_1 + \rho \lesseqgtr b_1 + l + s_1 - a$$

$$l \gtreqless \rho + a - s_1,$$

a relation which determines the necessary length of plate when s_1 is ascertained from the valve diagram.

On the other hand, it is unnecessary for the expansion plates to cut off later than the main valve, and for that grade of expansion they may reopen immediately, for the main valve will have closed the steam port. To secure this condition

$$s_2 = \rho$$

and the plates are then closest together. Hence, if the distance L has to be made as small as possible,

$$\text{L} \gtreqless l + \rho,$$

a relation which determines the minimum distance between the ports through the main valve. The ports may have to be sloped to get room enough for the expansion plates.

It is common to make the expansion eccentric radius r_2 equal to the main valve eccentric radius r_1. But a sharper cut off is obtained and the arrangement of the valve is made easier by taking

$$r_2 = 1\tfrac{1}{4} r_1.$$

158. Complete diagram for a Meyer gear.—Fig. 158 is a complete valve diagram for a Meyer valve gear. The three circles—crank circle, relative eccentric circle and main valve eccentric circle—are first drawn. Then with radius equal to the connecting-rod length the dead centre arcs at I and E. On the stroke line any desired points of cut off $\tfrac{1}{8}, \tfrac{1}{4}$. . . . are marked and arcs through these parallel to the dead-centre arcs cut the crank circle in positions of the crank for those grades of expansion.

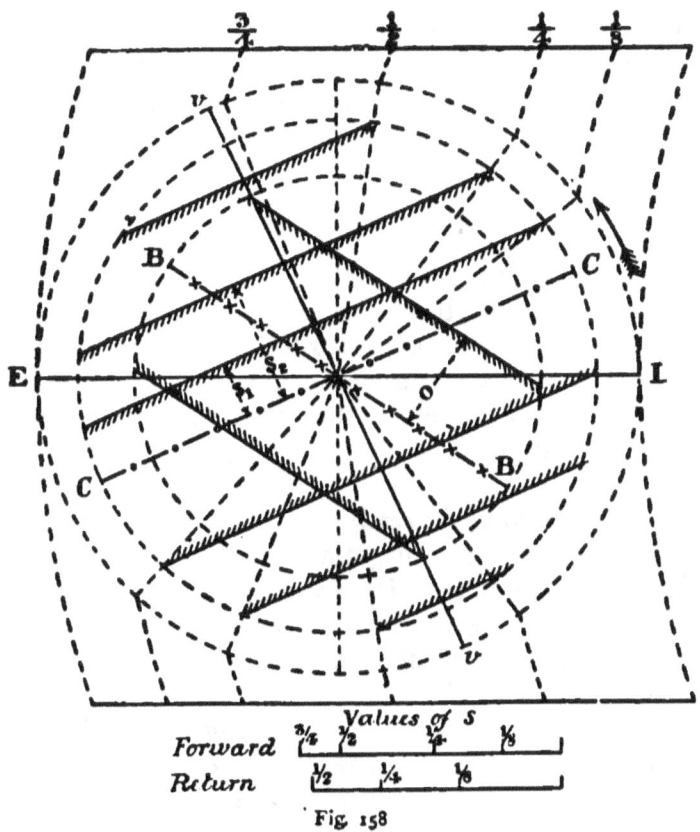

Fig. 158

Now draw BB, making the main valve angle of advance θ_1 with IE, and parallel to it the lap lines for the main valve. The crank positions at which the main valve closes the steam ports are given at vv. It will be seen that the main valve cuts off a little after $\tfrac{3}{4}$ stroke in the forward stroke and a little before it in the return stroke. The difference due to the obliquity of the connecting rod is here marked because a short connecting rod has been chosen.

Next draw CC, making the relative eccentric angle of advance ϕ with IE, reckoning in a direction reverse to the crank's motion. If the latest cut off of the expansion plates is to be that at which the main valve cuts off, CC should be at right angles to the position v of the crank at which the main valve cuts off. If the relative eccentric radius is chosen and its angle of advance fixed thus, then the radius and angle of advance of the expansion eccentric can be found, as in fig. 154.

At the points where the crank at the desired grades of cut off cuts the relative eccentric circle, draw lines parallel to CC. These mark off the values $s_1\ s_2\ \ldots$ to which the valve must be set for these grades of expansion. The little figure below gives the values of s for the forward and return stroke, and these are necessarily unequal. By setting the valve a little out of centre, the cut off may be equalised exactly for any given cut off and approximately for the others.

159. *Application of Zeuner's polar diagram to Meyer's valve gear.*—Zeuner's diagram may be used very conveniently in designing a Meyer valve gear. Two theorems additional to the one above for a single slide valve may be given first, in order to show how the valve circles for the relative eccentric are found.

160. *Theorem II. The polar locus of the relative travel of two valves moved by two eccentrics of different radii keyed at the same angle is a pair of circles.*—Let fig. 159 represent two slide valves (the valve rods are suppressed for simplicity) driven by two eccentrics, of_1, of_2 keyed at the same angle $90° + \theta$ with the crank ob. In diagram B take DD, making the angle of advance θ with YY, and with centres on DD and radii equal to half the eccentric radii draw the pairs of valve circles 1, 1, and 2, 2. Draw oqp parallel to the crank ob

Expansion Valves and Link Motions

In the position shown, valve 1 will have moved a distance $\xi_1 = oq$ to the left from its mid position and valve 2 a distance $\xi_2 = op$. Then the relative motion or relative travel of the valves is $\xi_2 - \xi_1 = pq$. Take $or = pq$ and join vp, tq, and draw rs perpendicular to orp. The angles vpo, tqo are right angles, being angles in a semicircle. Hence vp, tq, sr are parallel. But or by construction is equal to qp, and r is

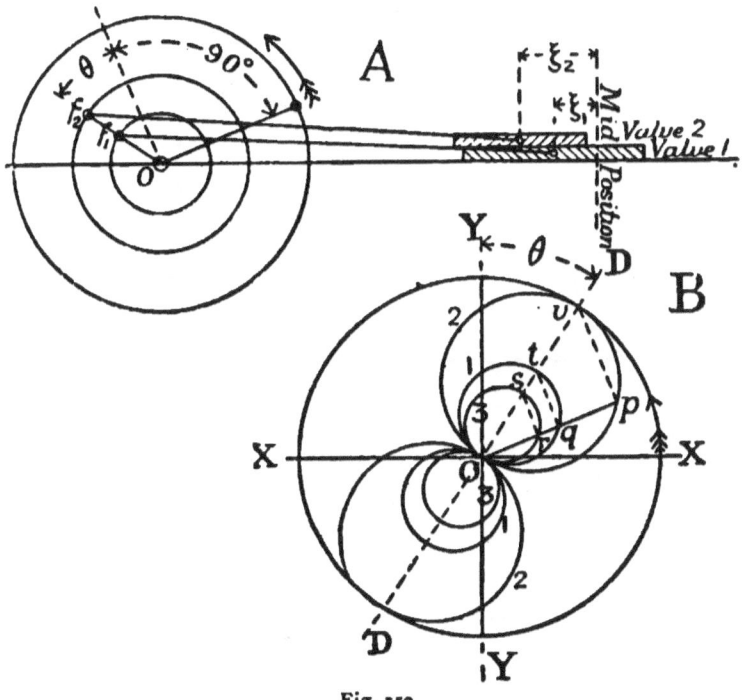

Fig. 159

always an apex of a right-angled triangle the hypothenuse of which is os. Hence the locus of r is the pair of circles marked 3. Consequently the relative travel of two valves, reckoned from their mid position, driven by eccentrics keyed at the same angle, is given by the intercept on the crank of a pair of valve circles which touch at o whose centres are on a line making an angle $90° - \theta$ with the initial position of the crank, and whose diameters are equal to the difference of the eccentric radii. The circles 3 3 are then the valve circles of the relative eccentric.

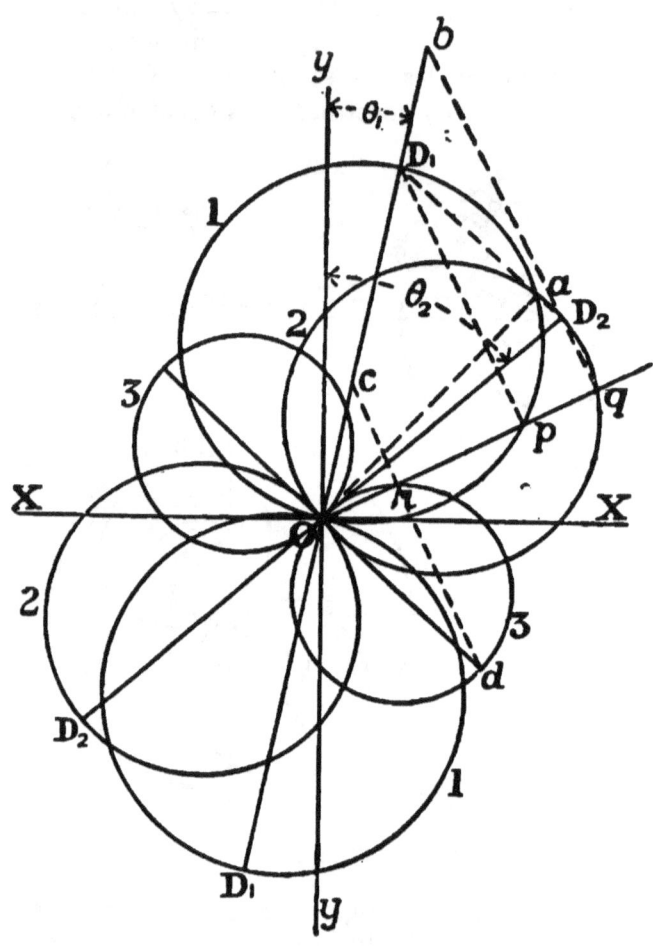

Fig. 160

161. Theorem III. *The polar locus of the relative travel of two valves driven by eccentrics with different angles of advance is a pair of circles.*—In fig. 160, let 1, 1 be the valve circles for a valve having the angle of advance θ_1, and

2 2 the valve circle for a valve having the angle of advance θ_2. D_1OD_1 and D_2OD_2 are the diameters of these circles. Let opq be any position of the crank. Then one valve will have travelled $\xi_1 = op$ from mid position; the other $\xi_2 = oq$. The relative travel is $\xi_2 - \xi_1 = pq$.

The three points D_1 a D_2 are in one straight line, the angles D_1 a o and D_2 a o being angles in semicircles. Draw $D_1 p$ and $D_2 q$ and produce the latter to meet oD_1 in b. These lines are perpendicular to opq since the angles at p and q are angles in semicircles. Hence they are parallel.

Take $or = pq$ and draw crd perpendicular to opq. This meets oD_1 in c and a line od parallel to $D_1 D_2$ in d. A circle 3 described on od will pass through r

or being equal to pq, oc is equal to $D_1 b$. But the angle $ocr = D_1 b D_2$ and the angle $cod = b D_1 D_2$. Hence the triangles cod and $b D_1 D_2$ are equal, and $od = D_1 D_2$.

Hence r lies on a circle 3 described on od as diameter, and od is equal and parallel to $D_1 D_2$, the line joining the ends of the diameters of the primary valve circles. Consequently, to draw the valve circles for the relative eccentric, take od equal and parallel to $D_1 D_2$, and describe on it the circles 3 3 touching at o. oa, drawn to the intersection of the primary valve circles, is a tangent to the valve circles for the relative eccentric; or, the intercept on the crank, is the relative travel of the two valves for that position of the crank.

162. *To draw a complete polar diagram for a Meyer gear.*—Take 1 E, fig. 161, to represent the length of stroke and draw the crank circle and dead centre arcs. On lines aa, bb, which are equal to 1 E, set off for the forward and return strokes the points of the stroke at which the expansion valve is to cut off steam, and find the corresponding crank positions by drawing arcs to the crank circle parallel to the dead-centre arcs.

Now draw the main valve circles 1 1, the eccentric radius being oD_1 and the angle of advance $AOD_1 = \theta_1$. With radius o $m = o$, the outside lap of main valve, draw the lap arcs. Then steam is cut off by the main valve in the crank positions o v, o v' which pass through the intersections of the valve circles and lap arcs. It will be seen that steam is cut off at

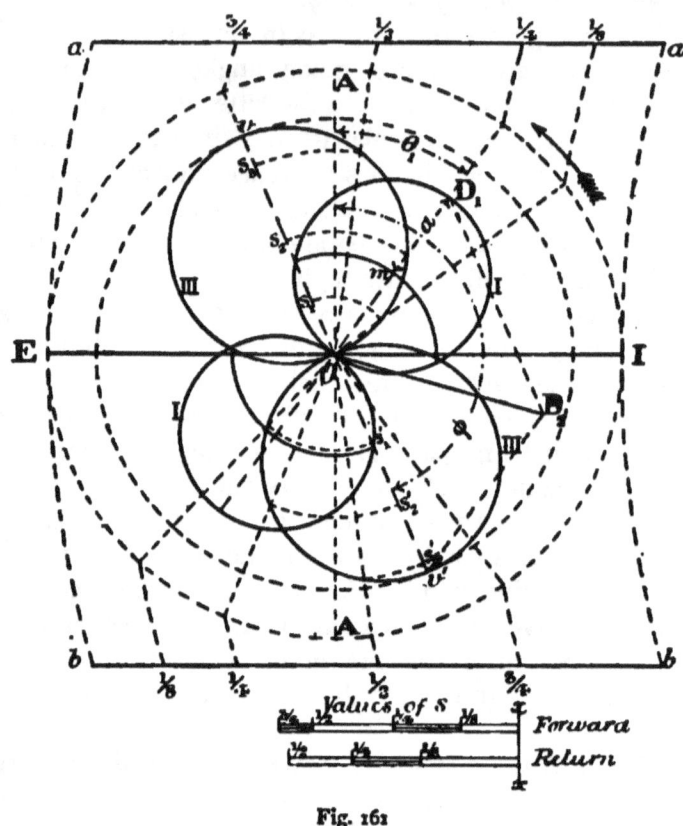

Fig. 161

$\frac{3}{4}$ stroke in the forward stroke and at about $\frac{5}{8}$ stroke in the return stroke.

If the expansion eccentric radius o $D_2 = r_2$ and its angle of advance, $\theta_2 = $ A O D_2 are given, complete the parallelogram O D_1 D_2 v'. Then o v' will be the radius ρ of the relative eccentric, and A O v' will be its angle of advance. This will not necessarily fall, as in fig. 161, on the crank position o v', at which the main valve cuts off. But it is convenient that it should do so, for then the latest cut off by the expansion valve coincides with that by the main valve, and the range of action of the gear is greatest. It is therefore more convenient to assume the relative eccentric $o\,v'$, determining the angle of advance by the crank position $o\,v'$, at which the main valve closes the steam port. Then the expansion eccentric $o\,D_2$ can be found by drawing the parallelogram $o\,D_1\,D_2\,v'$.

Suppose the relative eccentric radius $o\,v$ found or assumed, and draw the relative eccentric valve circles III, III. $A\,o\,v'$ reverse to the direction of motion is the angle of advance ϕ of the relative eccentric. Through the intersections of the relative valve circle and the crank positions at cut off, draw the dotted circles. Then $o\,s_1$, $o\,s_2$, ... are the values of s (fig. 157) for these grades of expansion. These values have been set off from $x\,x$ below. The inequalities in the values of s for any given cut off in the forward and return stroke are obvious. By adjusting the valve so that, say at $\frac{1}{8}$ cut off, s has the values found for both strokes, there will be less inequality of cut off for the other grades of expansion also.

Supposing the radius of relative eccentric and its angle of advance assumed, then the expansion eccentric is found thus. Draw $D_1\,D_2$ parallel to $o\,v$, and $v\,D_2$ parallel to $o\,D_1$. Then $o\,D_2$ is the radius of expansion eccentric and $A\,o\,D_2$ its angle of advance.

It may be pointed out in repetition that if the main valve just fully opens the steam port, the width of port a must be the distance $m\,D_1$. $o\,D_1$ is the radius r_1 of the main valve eccentric; $o\,D_2$ the radius r_2 of expansion eccentric; $o\,v$ the radius ρ of the relative eccentric. The length l of the expansion plates must not be less than $\rho + a - s^1$, where s_1 is the value of s for the earliest cut off. In the figure l must not be less than $s_1\,v + m\,D_1$. The distance L from centre to edge of steam port of main valve must be equal at least to $l + \rho$ or to $s_1\,v + m\,D_1 + o\,v$.

268 *Machine Design*

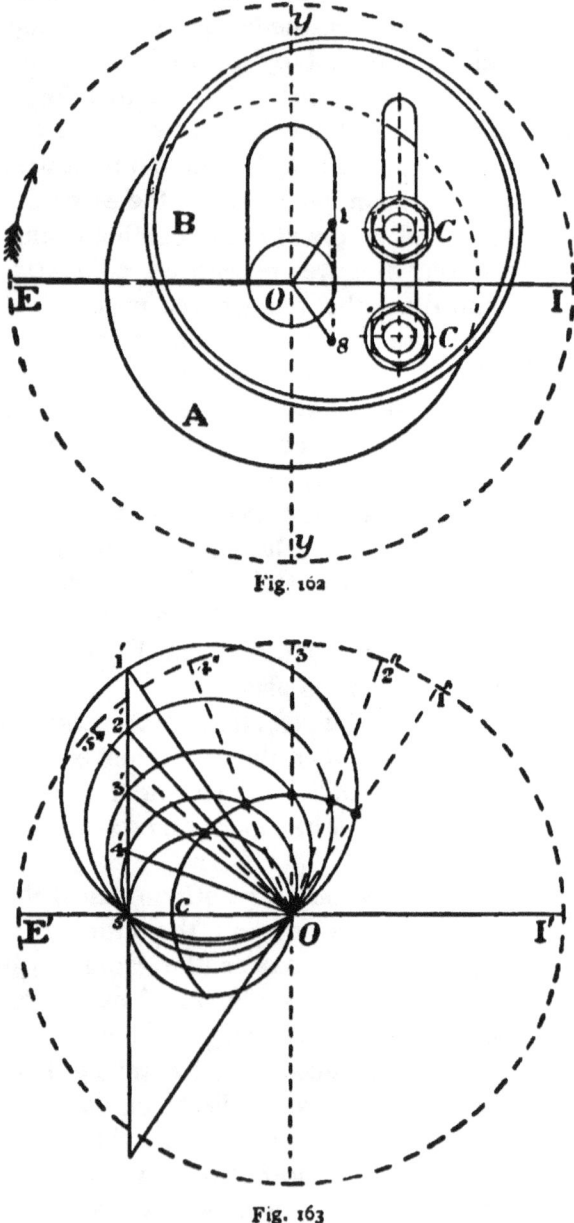

Fig. 162

Fig. 163

If a Meyer gear is used on an engine which reverses, the angle of advance of the expansion eccentric should be 90°, if the action is to be the same in forward and backward gear. Sometimes, however, it is necessary to be content with a good action of the expansion plates in forward gear and a more imperfect action in backward gear. Then the condition $\theta_2 = 90°$ is not imperative.

163. *Expansion by a movable eccentric.*—Suppose a plate A, fig. 162, keyed on the shaft and an eccentric cam B attached to this by bolts cc. If the cam is slotted as shown, it may be moved by slacking the bolts so that its centre travels along the straight line 1-8. In the position drawn the eccentric cam centre is at 1, the radius of eccentricity is o1, and the angle of advance yo1. But as the cam shifts the radius diminishes and the angle of advance increases.

Fig. 163 shows the valve diagrams for five positions of the cam to a larger scale. The outside lap circle has been drawn and the intersections of this with the valve circles determine the crank positions at cut off $1''$, $2''$, $3''$ . . . . for the chosen positions of the cam. If the path 1-8 of the eccentric centre is a straight line perpendicular to the crank the linear lead c5' is constant for all positions of the cam though the crank angle at which steam is admitted varies. When the angle of advance is greater than 90° the engine reverses.

There are mechanical difficulties in arranging for the shifting of the cam while the engine is running except in one important case. Of late powerful shaft governors have been introduced, keyed on and revolving with the crank shaft. It is possible to connect the shifting cam of an eccentric directly with one of these governors, so that the position of the cam and consequently the point of cut off of steam is absolutely controlled by the governor. The Turner-Hartnell governor is applied in this way. As the speed of the engine increases, the eccentric cam is shifted and steam cut off earlier. The path of the eccentric is not always a straight line. Then the linear amount of lead is not constant for different positions of the eccentric cam.

LINK MOTIONS

164. If an engine with a simple slide valve is required to reverse so that the crank shaft rotates at will either clockwise or counter-clockwise, means must be provided for placing the valve under the control of either of two eccentrics keyed at angles equal to $90° + \theta$ in both directions of rotation from the crank. An arrangement by which a single eccentric can be turned into these two positions is used on some marine engines. But it is more convenient generally to have two fixed eccentrics. Then the almost universal method of driving the slide valve is by a link connecting the two eccentric rod ends. The valve rod is connected with a block sliding on the link or in a slot formed in the link. By shifting the link, either eccentric rod can be brought opposite the slide block, and then the valve receives motion from that eccentric. Adopted at first merely for reversing, it was soon found that the link motion was a convenient expansion gear, for, if the slide block occupies a position intermediate to its extreme positions in the link, it gets a motion due to both eccentrics; its travel is reduced and cut off occurs earlier in the stroke. Exactly at the centre of the link is a neutral position, where the motion of the valve is too small to admit steam for motion in either direction.

The motion of a valve driven by a link cannot be quite easily and simply calculated except by approximate methods. Most commonly a very beautiful extension of the polar diagram, due to Zeuner, is used, and Zeuner's analysis has undoubtedly shown clearly the conditions to aim at in arranging a link motion. But there is a defect in Zeuner's method for students and practical engineers. The diagram is arrived at by a somewhat tedious algebraic analysis, in the course of which, for simplicity, various approximations are adopted. No doubt, if the link motion is well designed and of normal type, the approximations are legitimate and lead to no serious practical error. But it is difficult for students or practical engineers to satisfy themselves as to the error introduced in the approximations, and if the link motion is not of normal type, if the eccentric rods are short, the centre line of the link at a distance from the eccentric-rod ends, or the mode of suspension unusual, then the error introduced may no longer be unimportant. In any case, from the difficulty of the analysis, engineers are tempted simply to adopt the final construction arrived at by Zeuner as a mere rule of thumb, and they then feel they are working in the dark as to the accuracy of the method.

It is possible to arrive at an approximate solution of the link-motion problem in a very simple way, and one involving no tedious algebraical analysis. Suppose this approximation adopted first in roughly designing the valve motion. Then, by an easily understood graphic method, the exact motion of the valve can be determined, however abnormal the proportions of the gear. If the action is found to be imperfect, modifications can be introduced and the effect of these again determined. So an engineer can satisfy himself by a method clear and familiar, and involving no suppressions or assumptions, that the valve gear will act properly.

165. *Travel of a slide valve driven obliquely by an eccentric.*—Let x o x (fig. 164) be the line of stroke and y o y at right angles to it. Suppose o $a = r$ is the radius of an eccentric which drives b along the path o′ $b\,d$ parallel to x o x. If the figure is drawn for the crank at the dead point, y o a is the angle of advance θ of the eccentric. It is required to find the travel of the point b for any movement of the eccentrics. Join o b and let b o x $= \beta$. If o z is at right angles to o b, then z o a is $\theta + \beta$.

If the point b were driven by o a along the path o $b\,c$, the travel $b\,c = \xi$ for any movement of the eccentric would be found by drawing an ordinary polar valve circle, for an eccentric radius $oa = r$, the line of stroke being obc and the angle of advance $zoa = \theta + \beta$. The actual movement of b along $o'bd$ will be found very approximately by taking $od = oc$. Then since ocd is very nearly a right angle, $bd = bc/\cos\beta$. Hence the actual travel along $o'bd$ is very approximately $\zeta = \xi/\cos\beta$. In other words, the actual travel of a valve driven by an eccentric oa obliquely is the same as if it were driven by an eccentric $o'a'$ directly, the radius $o'a' = r_1 = r/\cos\beta$, and the angle of advance $vOa' = zoa = \theta + \beta$.

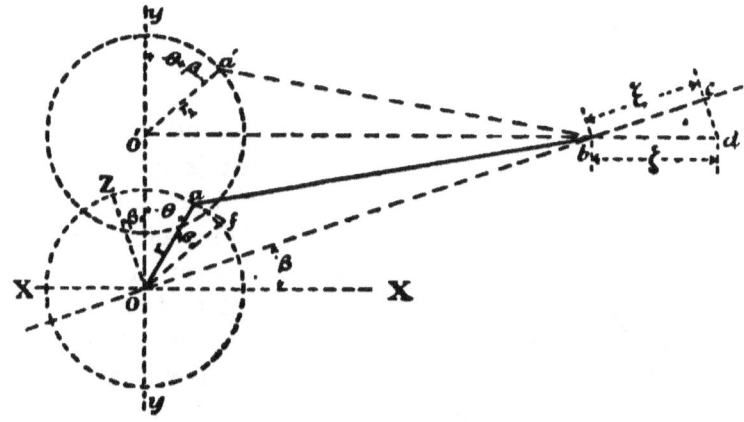

Fig. 164

Take $aof = \beta$, draw af at right angles to oa. Then $of = r/\cos\beta$ is the virtual eccentric radius and $vof = \theta + \beta$ the virtual angle of advance.

If β is measured from the line of stroke in the direction of instead of reverse to the direction of rotation, then β is negative and the virtual angle of advance is $\theta - \beta$.

166. *Stephenson's link motion.*—The link gear originally introduced by Stephenson is the simplest to construct and is the one most frequently used. The link (figs. 165, 166) is curved and concave to the crank shaft. of, ob are the forward and backward eccentrics for clockwise and counter-

clockwise rotation. *a d* is the link which is curved to a radius equal to the eccentric-rod length, or, more strictly, to a radius equal to the distance from the centre of an eccentric to the centre line of the link, measured along the centre line of the eccentric rod. If both eccentric radii are turned to the link side of the crank shaft, then the rods may have the positions shown in fig. 165, *open rods*, or that in fig. 166,

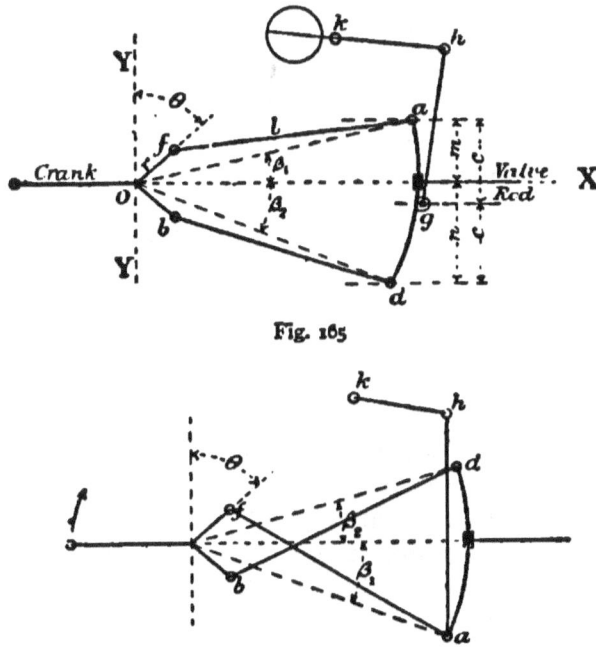

Fig. 165

Fig. 166

crossed rods. Either arrangement can be adopted. With open rods the linear lead of the valve increases as the link block moves towards the centre or neutral point of the link. With crossed rods the reverse is the case. Perhaps too much importance has been attached to this difference. Usually the eccentrics have equal angles of advance. By giving slightly unequal angles of advance the lead can be made nearly constant for forward running at the expense of greater variation in the backward running. If the link can be moved so that the block is precisely opposite an eccentric rod end, the half stroke of the valve is equal to the eccentric radius. Then the eccentric is of the smallest size and its friction is least. On the other hand, it must then usually be jointed to the link at a distance from its centre line, and this introduces some irregularity in the motion of the valve. If the eccentric is jointed to the link on its centre line, then for most forms of link the block cannot be brought opposite an eccentric-rod end. Then the eccentric radius must be greater than the half travel of the valve in full gear. The eccentrics, eccentric rods, and link form a four-bar chain, which is indefinitely deformable. To give a definite motion to the valve, one point of the link must be guided. This is usually done by suspending the link by a suspending link gh carried by a lever hk, which often carries a balance weight to balance the weight of the link and rods. By moving the lever kh the link block is brought to any part of the link. The lever is held in position in working by a catch, which drops into one of a set of notches.

The mode of suspension is important. The lever hk should be parallel to the line of stroke of the valve when the link block is at the neutral point of the link. When the valve is in mid position, a line through g perpendicular to the line of stroke should bisect the versed sine of the arc in which h moves. The lever hk should be as long as possible, usually not less than a quarter of the eccentric-rod length. Lastly, the suspending link hg should be as long as convenient. The object to aim at is that g should move as nearly as may be parallel to the line of stroke. There is least slip of the link block in the link if g is at the middle of the link. But the suspending rod is often attached to the lower end of the link to get a longer suspending link. The slip of the link block in the link is a serious practical evil, because it causes wear and consequent slackness of fit.

167. *Approximate designing of a Stephenson link motion.*

Equivalent eccentric for any position of the block in the link.
—If the ends of the link are supposed to move parallel to the line of stroke of the valve rod, then the method of § 167 may be used to find an eccentric which would give the valve a motion nearly identical with that which it receives from the two eccentrics driving the link.

In fig. 167 let o E be the crank at the dead point. $of = ob = r$, the two eccentric radii, here drawn with equal angles of advance $Y o f = Y o b = \theta$. Let the angles β_1 β_2, figs. 165, 166, be determined from a drawing of the link motion, which for convenience may generally be on a

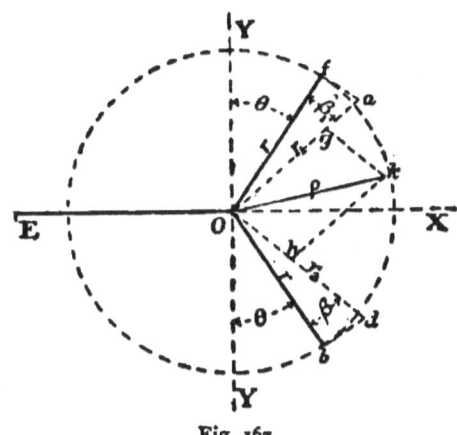

Fig. 167

smaller scale than the construction in fig. 167. Take $foa = \beta_1$ for open rods. For crossed rods the angle would be taken on the other side of of. Draw fa perpendicular to of. Then oa, as in fig. 164, is the eccentric which, driving the end of the link directly, would give it the same motion as of driving it obliquely, $oa = r_1 = r/\cos \beta_1$. Similarly find $od = r_2 = r/\cos \beta_2$ for the other eccentric.

Let m and n, figs. 165, 166, be the distances of the valve rod from the lines of stroke of the ends of the link, the whole length of link being $2c$. Then the motion given to the block by the forward eccentric will be less than that at the end of the link in the ratio $n/2c$, and that given by the backward eccentric in the ratio $m/2c$. Take $og = nr_1/2c$ and $oh = mr_2/2c$. Complete the parallelogram $ogkh$, then $ok = \rho$ is the equivalent eccentric. Valve circles drawn for an eccentric of radius ρ and angle of advance γok will give very approximately the motion of the valve.

If ad is joined and divided in the same ratio that the block divides the link, the point k is found even more simply.

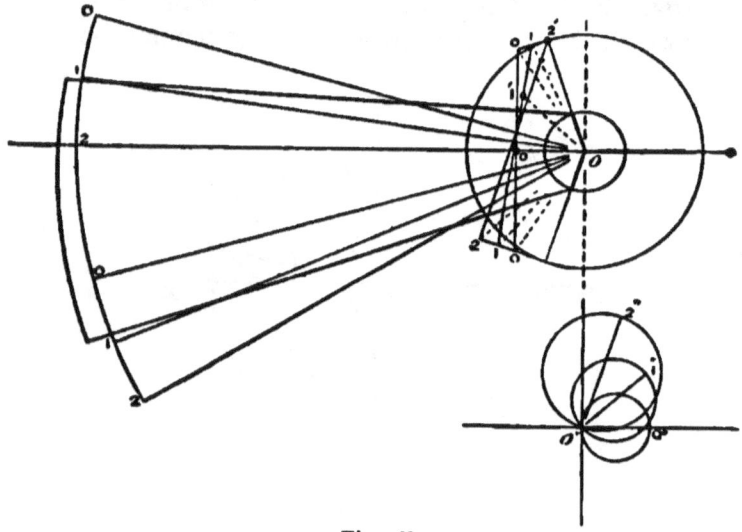

Fig. 168

Fig. 168 shows a link motion in three positions. $o\ o'$ $o\ 1'$, $o\ 2'$ are the three equivalent eccentrics determined as in §167. The lower figure gives the three corresponding Zeuner valve circles with diameters $o\ o'$, $o\ 1'$, $o\ 2'$. The smallest valve circle corresponds to the position when the slide block is at the neutral point. The largest valve circle is for full gear.

168. *Determination graphically of the exact travel of the valve for any crank position.*—It remains to indicate how, if a link gear has been designed provisionally, the movement

of the valve can be determined exactly, with a view to amendment if necessary. To obtain accuracy enough the movement of the valve must be determined full size or larger, and this makes it impossible to draw the whole link gear. A method is wanted not involving the drawing of the complete gear to so large a scale.

Let fig. 169 represent the gear to be examined, oc being the crank in any position, and of, ob being the corresponding positions of the forward and backward eccentrics. The position of the link will be determined if the positions of three points a, g, d can be determined. As to g there is no difficulty, for in all positions of the link g lies on the arc ge, struck from centre h with radius equal to the length of the suspending rod. The point a lies on an arc am struck from f with radius fa, and the point d on an arc dn struck from b with radius bd. The arcs am, dn, ge being drawn, a tracing of the link can be adjusted so that the points a, g, d fall on the director arcs, and then the intersection p of the centre line of the link with the line of stroke ox of the valve rod determines the position of the valve. By taking 8 or 12 positions of the eccentrics a corresponding number of positions of the valve can be determined by pricking through the tracing of the link. These can then be used to draw a valve ellipse showing completely the action of the valve for the given point of suspension h, and the movement for any other notch can be ascertained in the same way.

There is no great difficulty in drawing the link full size, but there would be a difficulty in drawing arcs with radius equal to the length of the eccentric rods. It remains to be seen how the director arcs am, nd can be determined without making use of the eccentric circle on the left of fig. 169.

Take $oo' = fa = bd$. With o' as centre draw a circle of radius equal to the eccentric radius, and draw $o'f'$, $o'b'$ parallel to of, ob; then f' and b' are points on the director arcs $af'm$ and $db'n$. Further, $b'b$ and $f'f$, parallel to oo'

will be radii of those arcs. Suppose an eccentric circle drawn and two simultaneous positions o, f' o b' of the eccentrics given. Let a templet rst (fig. 169, B) be prepared, by using a pear-wood curve, for instance, of the radius $l=fa=bd$, and having the side st in the direction of a radius. Placing this templet with ts coinciding with ff and $b'b$, the arcs am, dn can be drawn, and then with a tracing of the link the point p is determined.

The use of a templet in this way is due to MM. Coste

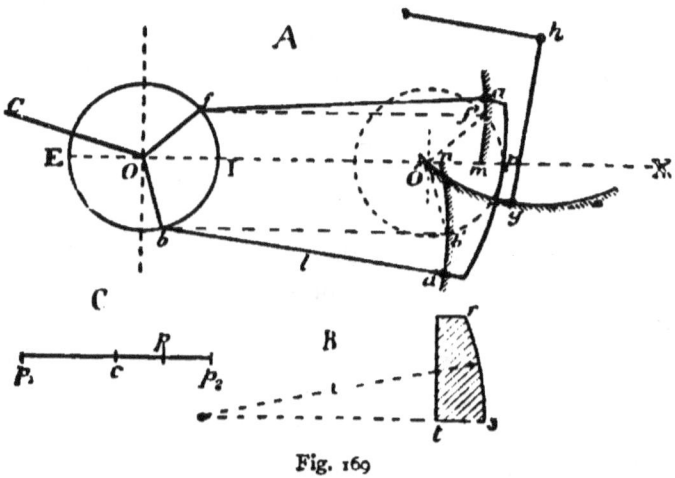

Fig. 169

and Maniquet. Although the process is cumbrous to describe, it is really a very easy one. Suppose that round the eccentric circle eight or twelve positions of f' are first marked and numbered, then eight or twelve corresponding positions of b' also numbered. The director arcs am, dn can then be all drawn and numbered, and these arcs suffice for all positions of the suspending link. Next one of the arcs, ge, can be drawn and the link, tracing used to mark off twelve valve positions for that notch. If the positions of a and d are pricked through (or, more strictly, the ends of the link centre line) twelve points are determined in the peculiar curves described by the ends of the link, and termed *slip curves*.

A new position of the director arc ge can then be taken and a new set of values of the valve travel determined, the same director arcs for the points such as a and d being used as for the previous notch.

To set off the valve travels thus determined properly on an ellipse diagram, the true position of the centre of the cylinder face must be known. Usually the valve is set to equal linear leads for one notch; for instance, for full gear forward. That is, the travel of the valve reckoned from centre of cylinder face is $o + e$ with the crank at the interior and exterior dead points for that notch. Let p_1, p_2, fig. 169, c, be two positions of the point p for the given notch and for the crank at the dead points. Bisect $p_1 p_2$ in c. Then c is the true centre of the cylinder face relatively to the position p of the valve, determined by the construction, for that and all the other notches. $cp_1 = cp_2 = o + e$ for the notch at which the leads are equal. If p is any other position of p for any notch, cp is the travel of the valve relatively to the centre of the cylinder face.

169. *Example of a link motion valve gear.*—It may be useful as a guide in designing link motions to give some details of a carefully designed gear for a locomotive engine. The following data are from a gear designed by Mr. William Adams as a standard for outside cylinder engines on the North London Railway.[1]

The engine has cylinders 17 ins. diam. and 24 ins. stroke. The steam ports are $14\frac{1}{2} \times 1\frac{1}{4}$ ins., or 18·1 sq. in. in area. This is a little less than 1/12th of the piston area. The exhaust port is $14\frac{1}{2} \times 3\frac{1}{2}$ ins. The bars are 1 in. wide. There is no inside lap and the outside lap is 1 in. The area of the back of the valve is about 140 sq. ins., so that the frictional resistance to sliding, calculated in the ordinary way, might be taken at about 1,680 lbs., and this is the

[1] 'Engineering,' vol. viii. p. 112.

Table A.

Piston stroke, 24 inches; radius of eccentric, 3⅞ inches; lap of valve, 1 inch.

Notch	Travel of Valve	Lead		Maximum opening of Port		Fraction of Stroke at which steam is cut off		Fraction of Stroke at release		Maximum Slip of Block in Link
		Front Port	Back Port	Front Port	Back Port	Stroke towards Crank	Stroke from Crank	Stroke towards Crank	Stroke from Crank	
Forward	in.	in.	in.	in.	in.					in.
1st .	2·56	·28	·28	·28	·28	·11	·11	·58	·58	·12
2nd .	2·62	·28	·28	·31	·31	·22	·23	·65	·65	·12
3rd .	2·74	·28	·25	·37	·37	·31	·32	·72	·71	·19
4th .	2·91	·25	·25	·44	·47	·40	·41	·77	·76	·25
5th .	3·06	·25	·22	·5	·56	·48	·49	·82	·80	·31
6th .	3·31	·22	·19	·62	·69	·57	·57	·86	·84	·35
7th .	3·44	·22	·19	·72	·72	·64	·61	·88	·86	·37
8th .	3·81	·19	·16	·84	·92	·70	·67	·90	·88	·44
9th .	4·09	·19	·12	·97	1·12	·75	·71	·92	·90	·50
10th .	4·34	·16	·09	1·09	1·25	·79	·75	·94	·92	·56
Backward										
1st .	2·56	·28	·25	·28	·28	·11	·11	·58	·57	·03
2nd .	2·62	·28	·25	·31	·31	·23	·23	·65	·65	nil
3rd .	2·76	·28	·25	·37	·39	·31	·32	·73	·72	·03
4th .	2·91	·25	·22	·44	·47	·40	·41	·78	·77	·09
5th .	3·09	·22	·22	·53	·56	·48	·49	·82	·81	·12
6th .	3·31	·22	·19	·62	·69	·57	·57	·86	·85	·19
7th .	3·56	·19	·19	·75	·81	·64	·64	·89	·88	·25
8th .	3·84	·16	·16	·87	·97	·70	·69	·90	·90	·31
9th .	4·12	·12	·16	1·00	1·12	·75	·73	·92	·92	·37
10th .	4·37	·09	·12	1·12	1·25	·79	·78	·94	·94	·44

straining force which the parts of the link motion have to overcome. The eccentrics are 15¼ ins. diameter with straps 2⅜ ins. wide. At 40 miles per hour the engine would make about 210 revolutions per minute. Hence the width of the eccentric is about

$$b = \frac{PN}{148,000}.$$

TABLE B.

Stroke of piston, 24 inches; radius of eccentric, 3¼ inches; lap, 1 inch. Set with all rods 5/64ths inch short. 5/32nds more opening in front port than back.

Notch	Travel of Valve	Lead		Maximum opening		Fraction of Stroke at cut off		Fraction of Stroke at release		Maximum Slip of Block Link
		Front Port	Back Port	Front Port	Back Port	Stroke towards Crank	Stroke from Crank	Stroke towards Crank	Stroke from Crank	
	in.	in.	in.	in.	in.					in.
Forward										
1st	2·53	·34	·16	·34	·19	·19	·12	·61	·55	·12
2nd	2·56	·34	·16	·34	·22	·26	·19	·69	·61	·19
3rd	2·72	·34	·16	·44	·28	·35	·28	·75	·68	·25
4th	2·87	·31	·12	·50	·37	·44	·38	·80	·73	·28
5th	3·06	·31	·12	·56	·50	·52	·46	·84	·78	·31
6th	3·31	·31	·09	·69	·62	·60	·53	·88	·81	·37
7th	3·53	·28	·09	·78	·75	·67	·59	·90	·84	·44
8th	3·78	·28	·06	·91	·87	·72	·65	·92	·86	·50
9th	4·03	·25	·03	1·03	1·03	·77	·69	·93	·89	·56
10th	4·35	·22	—	1·16	1·19	·80	·73	·95	·91	·57
Backward										
1st	2·50	·34	·16	·34	·16	·19	·13	·61	·55	·03
2nd	2·56	·34	·16	·34	·22	·26	·19	·68	·61	—
3rd	2·72	·34	·16	·44	·28	·34	·28	·74	·69	·06
4th	2·87	·31	·16	·50	·37	·43	·38	·80	·74	·09
5th	3·06	·31	·12	·56	·50	·52	·46	·84	·78	·12
6th	3·31	·28	·12	·69	·62	·59	·54	·87	·82	·19
7th	3·53	·25	·09	·78	·75	·66	·60	·90	·86	·25
8th	3·77	·22	·06	·91	·86	·71	·67	·92	·89	·31
9th	4·03	·19	·06	1·03	1·03	·76	·72	·93	·91	·44
10th	4·38	·16	·03	1·19	1·19	·80	·76	·95	·93	·50

Comparing this with the rule on p. 93, it will be seen that a smaller bearing surface is allowed than the rule there given provides for. It is true also, however, that the ordinary friction of the slide valve is probably little more than half that calculated above, that an engine is not always running at so high a speed, and that even in full gear the leverage of the link reduces the thrust on the eccentric a little below the value assumed above. The link is 18 ins. long, and the slide block $4\frac{1}{2}$ ins. long by $2\frac{3}{4}$ ins. wide. This gives an area of surface to the block of 12 sq. ins., so that the maximum pressure on the wearing surface of the block is about 140 lbs. per sq. in.

The preceding tables show the action of the link gear. Table A shows the most equable distribution of steam which could be obtained by the gear; Table B shows the action of the gear, with the setting actually adopted. It was found in running that the action was better with the setting shown in Table B. Mr. Adams states that the setting in Table B is arrived at in this way: The reversing lever is placed in mid gear and the crank at half-stroke. The length of the eccentric rods is then adjusted so that the valve closes the ports perfectly when the crank is on either the top or bottom centre.

CHAPTER XII

LUBRICATORS

170. The amount of the frictional resistance of machine parts which slide on each other depends on the smoothness of the surfaces and on their lubrication. A lubricant is a substance which, interposed between the rubbing surfaces, reduces the friction. The diminution of friction diminishes the work wasted, the wear of the rubbing parts, and the amount of heat developed. Surfaces which run perfectly cool and well, if properly lubricated, heat and even seize if the lubrication fails. Seizing is the cohering of the parts with force great enough to cause fracture or stoppage of the machine.

An efficient lubricant should possess the following qualities : (*a*) It should wet the rubbing surfaces. (*b*) It must not evaporate or decompose while in use. (*c*) At the temperature at which it is employed it should have enough, and only enough, viscosity to remain between the surfaces. (*d*) It must contain no acids or other constituents capable of acting on the rubbing surfaces. (*e*) It must be free from grit or other foreign matter.

Air or water are good lubricants when the velocity is great enough to carry in a layer between the rubbing surfaces. For instance, water is the lubricant for propeller shaft bearings in the stern tube, the bearing surfaces being strips of lignum vitæ. (See Part I. p. 265.)

Lubricants are sometimes solid at ordinary temperatures,

as tallow or railway grease, more commonly fluid, as vegetable, animal, or mineral oils. Metaline, plumbago, and some other materials are used without lubricants, and act themselves as lubricants.

Of vegetable oils, olive, palm, rape, and others are used. Of animal oils, sperm is one of the very best, but lard, neat's foot, seal, and other oils are used. Of mineral oils, some are derived from the distillation of shale, and others from petroleum wells. Heavy mineral oils have now largely superseded all other oils for purposes of lubrication. Railway grease, which is a solid or semi-fluid lubricant, is a

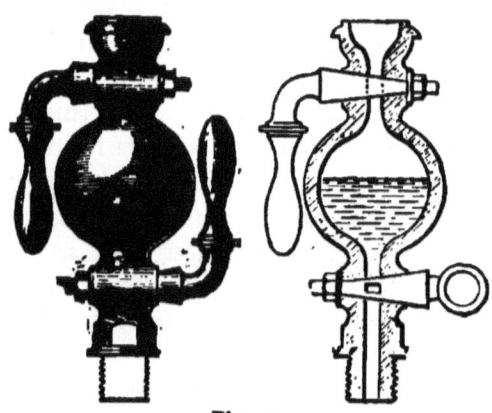

Fig. 170

mixture of tallow, palm oil, water, and a portion of caustic soda.

To provide for the proper lubrication of rubbing parts, a reservoir of the lubricant must be provided, so arranged, if possible, that it delivers the lubricant continuously or at regular short intervals in small quantities. The lubricant flows through an aperture to the rubbing surfaces, one of which is generally provided with channels for its suitable distribution. Lastly, in many cases a vessel must be provided to catch the lubricant which flows off the rubbing surfaces after it has done its work.

171. *Cup lubricators*.—The simplest lubricator is a cup which can be filled from time to time, and which has a cock, by turning which the lubricant is permitted to flow to the rubbing surfaces.

Fig. 170 shows the ordinary form of steam cylinder oil-cup or tallow-cup. It is intended to serve for introducing oil to the cylinder without causing an escape of steam. It consists of a vessel having two cocks. Closing the lower one and opening the upper one, it can be filled; closing the upper and opening the lower, the oil or melted tallow is admitted to the cylinder. It is often placed directly on the steam cylinder, sometimes on the steam pipe, where the rapid current of steam carries forward the oil to the working parts.

172. *Displacement lubricator.*—Fig. 171 shows the displacement lubricator invented by Ramsbottom. The steam, condensing on the surface of the oil, forms a drop which sinks down through the oil and displaces a small quantity of oil, which then flows down the bent pipe into the cylinder. The plug at the top serves for filling the cistern, that at bottom for removing the condensed water. Numerous forms of sight feed lubricators have been introduced. These are displacement lubricators, the oil drop ascending in a glass gauge tube.

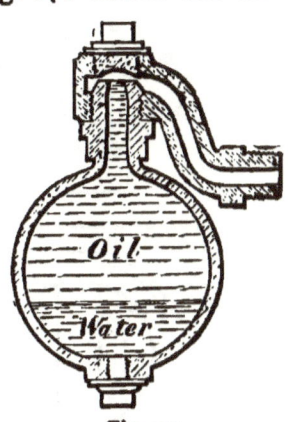

Fig. 171

173. *Siphon lubricator.*—Fig. 172 shows an ordinary siphon lubricator for oil. This has a tube rising above the surface of the oil in which a cotton wick is placed. The oil is slowly siphoned by the capillary action of the wick and drops on to the bearing. Various arrangements are adopted for closing the reservoir to keep out dirt. Sometimes there is a hinged or screwed cap. In the example shown there is a rotating plate with a hole. By rotating the

plate this hole can be brought over a hole in a lower plate or over a blank part of the lower plate. The amount of cotton wick necessary to supply sufficient oil is ascertained by trial.

174. *Needle lubricator.*—Fig. 173 shows Lieuvain's needle lubricator, oftenest used for the bearings of shafting.

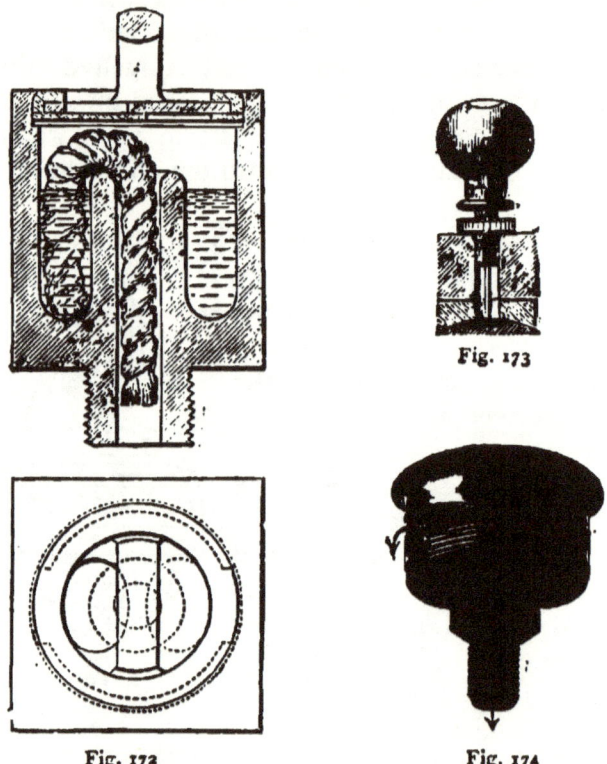

Fig. 172 Fig. 173 Fig. 174

It consists of a glass reservoir having a wooden plug. This is filled and inverted over the bearing. A wire needle or pin passes through the plug, loosely, and rests on the journal. When the shaft is running, the vibration of the needle causes a slow descent of the oil. When the shaft is at rest, the capillary attraction stops the flow of the oil. The rate of supply of oil may be regulated by making the needle thicker or thinner.

175. *Stauffer lubricator.*—Fig. 174 shows another kind of lubricator, in which a semi-fluid or grease lubricant is used. The cap can be screwed down on a fine pitched screw. This forces the lubricant down the tube to the bearing. The tube may even be of considerable length if convenient, the grease flowing under pressure like a fluid.

www.ingramcontent.com/pod-product-compliance
Lightning Source LLC
Chambersburg PA
CBHW032048230426
43672CB00009B/1515